W9-CEG-629

Dynamical systems and fractals

Computer graphics experiments
in Pascal

© Becker / Dörfler

Dynamical systems and fractals

Computer graphics experiments in Pascal

Karl-Heinz Becker
Michael Dörfler

Translated by Ian Stewart

1990

OCT

ELMHURST COLLEGE LIBRARY

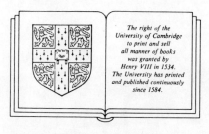

The right of the
University of Cambridge
to print and sell
all manner of books
was granted by
Henry VIII in 1534.
The University has printed
and published continuously
since 1584.

CAMBRIDGE UNIVERSITY PRESS
Cambridge
New York   Port Chester   Melbourne   Sydney

Published by the Press Syndicate of the University of Cambridge
The Pitt Building, Trumpington Street, Cambridge CB2 1RP
40 West 20th Street, New York, NY 10011, USA
10 Stamford Road, Oakleigh, Melbourne 3166, Australia

Originally published in German as *Computergrafische Experimente mit Pascal: Chaos und Ordnung in Dynamischen Systemen* by Friedr. Vieweg & Sohn, Braunschweig 1986, second edition 1988, and © Friedr. Vieweg & Sohn, Verlagsgesellschaft mbH, Braunschweig 1986, 1988

First published in English 1989
Reprinted 1990

English translation © Cambridge University Press 1989

Printed in Great Britain at the University Press, Cambridge

Library of Congress cataloguing in publication data available

British Library cataloguing in publication data
Becker, Karl-Heinze
  Dynamical systems and fractals
  1. Mathematics. Applications of computer graphics
  I. Title.   II. Dörfler, Michael   III.
  Computergrafische Experimente mit Pascal.   *English*
  510'.28'566

ISBN 0 521 36025 0 hard covers
ISBN 0 521 36910 X paperback

Contents

Foreword

New Directions in Computer Graphics:
Experimental Mathematics

As a mathematician one is accustomed to many things. Hardly any other academics encounter as much prejudice as we do. To most people, mathematics is the most colourless of all school subjects - incomprehensible, boring, or just terribly dry. And presumably, we mathematicians must be the same, or at least somewhat strange. We deal with a subject that (as everyone knows) is actually complete. Can there still be anything left to find out? And if yes, then surely it must be totally uninteresting, or even superfluous.

Thus it is for us quite unaccustomed that our work should so suddenly be confronted with so much public interest. In a way, a star has risen on the horizon of scientific knowledge, that everyone sees in their path.

Experimental mathematics, a child of our 'Computer Age', allows us glimpses into the world of numbers that are breathtaking, not just to mathematicians. Abstract concepts, until recently known only to specialists - for example Feigenbaum diagrams or Julia sets - are becoming vivid objects, which even renew the motivation of students. Beauty and mathematics: they belong together visibly, and not just in the eyes of mathematicians.

Experimental mathematics: that sounds almost like a self-contradiction! Mathematics is supposed to be founded on purely abstract, logically provable relationships. Experiments seem to have no place here. But in reality, mathematicians, by nature, have always experimented: with pencil and paper, or whatever equivalent was available. Even the relationship $a^2+b^2=c^2$, well-known to all school pupils, for the sides of a right-angled triangle, didn't just fall into Pythagoras' lap out of the blue. The proof of this equation came after knowledge of many examples. The working out of examples is a typical part of mathematical work. Intuition develops from examples. Conjectures are formed, and perhaps afterwards a provable relationship is discerned. But it may also demonstrate that a conjecture was wrong: a single counter-example suffices.

Computers and computer graphics have lent a new quality to the working out of examples. The enormous calculating power of modern computers makes it possible to study problems that could never be assaulted with pencil and paper. This results in gigantic data sets, which describe the results of the particular calculation. Computer graphics enable us to handle these data sets: they become visible. And so, we are currently gaining insights into mathematical structures of such infinite complexity that we could not even have dreamed of it until recently.

Some years ago the Institute for Dynamical Systems of the University of Bremen was able to begin the installation of an extensive computer laboratory, enabling its members to carry out far more complicated mathematical experiments.   Complex dynamical systems are studied here; in particular mathematical models of changing or self-modifying systems that arise from physics, chemistry, or biology (planetary orbits, chemical reactions, or population development).   In 1983 one of the Institute's research groups concerned itself with so-called *Julia sets*.   The bizarre beauty of these objects lent wings to fantasy, and suddenly was born the idea of displaying the resulting pictures as a public exhibition.

Such a step down from the 'ivory tower' of science, is of course not easy. Nevertheless, the stone began to roll.   The action group 'Bremen and its University', as well as the generous support of Bremen Savings Bank, ultimately made it possible: in January 1984 the exhibition *Harmony in Chaos and Cosmos* opened in the large bank lobby. After the hectic preparation for the exhibition, and the last-minute completion of a programme catalogue, we now thought we could dot the i's and cross the last t's. But something different happened: ever louder became the cry to present the results of our experiments outside Bremen, too.   And so, within a few months, the almost completely new exhibition *Morphology of Complex Boundaries* took shape.   Its journey through many universities and German institutes began in the Max Planck Institute for Biophysical Chemistry (Göttingen) and the Max Planck Institute for Mathematics (in Bonn Savings Bank).

An avalanche had broken loose.   The boundaries within which we were able to present our experiments and the theory of dynamical systems became ever wider.   Even in (for us) completely unaccustomed media, such as the magazine *Geo* on ZDF television, word was spread.   Finally, even the Goethe Institute opted for a world-wide exhibition of our computer graphics.   So we began a third time (which is everyone's right, as they say in Bremen), equipped with fairly extensive experience.   Graphics, which had become for us a bit too brightly coloured, were worked over once more. Naturally, the results of our latest experiments were added as well.   The première was celebrated in May 1985 in the 'Böttcherstrasse Gallery'.   The exhibition *Schönheit im Chaos/Frontiers of Chaos* has been travelling throughout the world ever since, and is constantly booked.   Mostly, it is shown in natural science museums.

It's no wonder that every day we receive many enquiries about computer graphics, exhibition catalogues (which by the way were all sold out) and even programming instructions for the experiments. Naturally, one can't answer all enquiries personally.   But what are books for? *The Beauty of Fractals*, that is to say the book about the exhibition, became a prizewinner and the greatest success of the scientific publishing company Springer-Verlag. Experts can enlighten themselves over the technical details in *The Science of Fractal Images*; and with *The Game of Fractal Images* lucky Macintosh II owners, even without any further knowledge, can boot up their computers and go on a journey of discovery at once.   But what about all the many home computer fans, who themselves like to program, and thus would like simple, but exact. information?   The book lying in front of you by Karl-Heinz Becker and Michael Dörfler fills a gap that has too long been open.

The two authors of this book became aware of our experiments in 1984, and through our exhibitions have taken wing with their own experiments. After didactic preparation they now provide, in this book, a quasi-experimental introduction to our field of research. A veritable kaleidoscope is laid out: dynamical systems are introduced, bifurcation diagrams are computed, chaos is produced, Julia sets unfold, and over it all looms the 'Gingerbread Man' (the nickname for the Mandelbrot set). For all of these, there are innumerable experiments, some of which enable us to create fantastic computer graphics for ourselves. Naturally, a lot of mathematical theory lies behind it all, and is needed to understand the problems in full detail. But in order to experiment oneself (even if in perhaps not quite as streetwise a fashion as a mathematician) the theory is luckily not essential. And so every home computer fan can easily enjoy the astonishing results of his or her experiments. But perhaps one or the other of these will let themselves get really curious. Now that person can be helped, for that is why it exists: the study of mathematics.

But next, our research group wishes you lots of fun studying this book, and great success in your own experiments. And please, be patient: a home computer is no 'express train' (or, more accurately, no supercomputer). Consequently some of the experiments may tax the 'little ones' quite nicely. Sometimes, we also have the same problems in our computer laboratory. But we console ourselves: as always, next year there will be a newer, faster, and simultaneously cheaper computer. Maybe even for Christmas... but please with colour graphics, because then the fun *really* starts.

Research Group in Complex Dynamics
University of Bremen
                      Hartmut Jürgens

Preface to the German Edition

Today the 'theory of complex dynamical systems' is often referred to as a revolution, illuminating all of science. Computer-graphical methods and experiments today define the methodology of a new branch of mathematics: 'experimental mathematics'. Its content is above all the theory of complex dynamical systems. 'Experimental' here refers primarily to computers and computer graphics. In contrast to the experiments are 'mathematical cross-connections', analysed with the aid of computers, whose examples were discovered using computer-graphical methods. The mysterious structure of these computer graphics conceals secrets which still remain unknown, and lie at the frontiers of thought in several areas of science. If what we now know amounts to a revolution, then we must expect further revolutions to occur.

- The groundwork must therefore be prepared, and
- people must be found who can communicate the new knowledge.

We believe that the current favourable research situation has been created by the growing power and cheapness of computers. More and more they are being used as research tools. But science's achievement has always been to do what can be done. Here we should mention the name of Benoі§t B. Mandelbrot, a scientific outsider who worked for many years to develop the fundamental mathematical concept of a fractal and to bring it to life.

Other research teams have developed special graphical techniques. At the University of Bremen fruitful interaction of mathematicians and physicists has led to results which have been presented to a wide public. In this context the unprecedented popular writings of the group working under Professors Heinz–Otto Peitgen and Peter H. Richter must be mentioned. They brought computer graphics to an interested public in many fantastic exhibitions. The questions formulated were explained non-technically in the accompanying programmes and exhibition catalogues and were thus made accessible to laymen. They recognised a further challenge, to emerge from the 'Ivory Tower' of science, so that scientific reports and congresses were arranged not only in the university. More broadly, the research group presented its results in the magazine *Geo*, on ZDF television programmes, and in worldwide exhibitions arranged by the Goethe Institute. We know of no other instance where the bridge from the foremost frontier of research to a wide lay public has been built in such a short time. In our own way we hope to extend that effort in this book. We hope, while dealing with the discoveries of the research group, to open for many readers the path to their own experiments. Perhaps in this way we can lead them towards a deeper understanding of the problems connected with mathematical feedback.

Our book is intended for everyone who has a computer system at their disposal and who enjoys experimenting with computer graphics. The necessary mathematical formulas are so simple that they can easily be understood or used in simple ways. The reader will rapidly be brought into contact with a frontier of today's scientific research, in which hardly any insight would be possible without the use of computer systems and graphical data processing.

This book divides into two main parts. In the first part (Chapters 1-10), the reader is introduced to interesting problems and sometimes a solution in the form of a program fragment. A large number of exercises lead to individual experimental work and independent study. The first part closes with a survey of 'possible' applications of this new theory.

In the second part (from Chapter 11 onwards) the modular concept of our program fragments is introduced in connection with selected problem solutions. In particular, readers who have never before worked with Pascal will find in Chapter 11 - and indeed throughout the entire book - a great number of program fragments, with whose aid independent computer experimentation can be carried out. Chapter 12 provides reference programs and special tips for dealing with graphics in different operating systems and programming languages. The contents apply to MS-DOS systems with Turbo Pascal and UNIX 4.2 BSD systems, with hints on Berkeley Pascal and C. Further example programs, which show how the graphics routines fit together, are given for Macintosh systems (Turbo Pascal, Lightspeed Pascal, Lightspeed C), the Atari (ST Pascal Plus), the Apple IIe (UCSD Pascal), and the Apple IIGS (TML Pascal).

We are grateful to the Bremen research group and the Vieweg Company for extensive advice and assistance. And, not least, to our readers. Your letters and hints have convinced us to rewrite the first edition so much that the result is virtually a new book - which, we hope, is more beautiful, better, more detailed, and has many new ideas for computer graphics experiments.

Bremen   *Karl-Heinz Becker*   *Michael Dörfler*

1 Researchers Discover Chaos

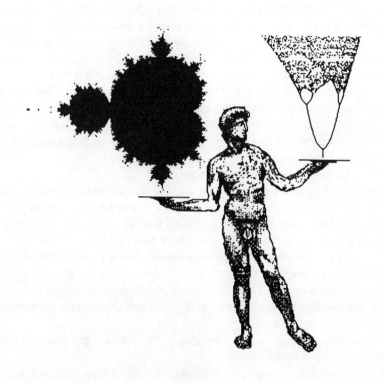

The story which today so fascinates researchers, and which is associated with chaos theory and experimental mathematics, came to our attention around 1983 in Bremen. At that time a research group in dynamical systems under the leadership of Professors Peitgen and Richter was founded at Bremen University. This starting-point led to a collaboration lasting many years with members of the Computer Graphics Laboratory at the University of Utah in the USA.

Equipped with a variety of research expertise, the research group began to install its own computer graphics laboratory. In January and February of 1984 they made their results public. These results were startling and caused a great sensation. For what they exhibited was beautiful, coloured computer graphics reminiscent of artistic paintings. The first exhibition, *Harmony in Chaos and Cosmos*, was followed by the exhibition *Morphology of Complex Frontiers*. With the next exhibition the results became internationally known. In 1985 and 1986, under the title *Frontiers of Chaos* and with assistance from the Goethe Institute, this third exhibition was shown in the UK and the USA. Since then the computer graphics have appeared in many magazines and on television, a witches' brew of computer-graphic simulations of dynamical systems.

What is so stimulating about it?

Why did these pictures cause so great a sensation?

We think that these new directions in research are fascinating on several grounds. It seems that we are observing a ' celestial conjunction' – a conjunction as brilliant as that which occurs when Jupiter and Saturn pass close together in the sky, something that happens only once a century. Similar events have happened from time to time in the history of science. When new theories overturn or change previous knowledge, we speak of a *paradigm change.* [1]

The implications of such a paradigm change are influenced by science and society. We think that may also be the case here. At any rate, from the scientific viewpoint, this much is clear:

- A new theory, the so-called *chaos theory*, has shattered the scientific world-view. We will discuss it shortly.
- New techniques are changing the traditional methods of work of mathematics and lead to the concept of *experimental mathematics*.

For centuries mathematicians have stuck to their traditional tools and methods such as paper, pen, and simple calculating machines, so that the typical means of progress in mathematics have been proofs and logical deductions. Now for the first time some mathematicians are working like engineers and physicists. The mathematical problem under investigation is planned and carried out like an experiment. The experimental apparatus for this investigatory mathematics is the computer. Without it, research in this field would be impossible. The mathematical processes that we wish to understand are

---

[1] Paradigm = 'example'. By a paradigm we mean a basic point of view, a fundamental unstated assumption, a dogma, through which scientists direct their investigations.

visualised in the form of computer graphics. From the graphics we draw conclusions about the mathematics. The outcome is changed and improved, the experiment carried out with the new data. And the cycle starts anew.

- Two previously separate disciplines, mathematics and computer graphics, are growing together to create something qualitatively new.

Even here a further connection with the experimental method of the physicist can be seen. In physics, bubble-chambers and semiconductor detectors are instruments for visualising the microscopically small processes of nuclear physics. Thus these processes become representable and accessible to experience. Computer graphics, in the area of dynamical systems, are similar to bubble-chamber photographs, making dynamical processes visible.

Above all, this direction of research seems to us to have social significance:

- The 'ivory tower' of science is becoming transparent.

In this connection you must realise that the research group is interdisciplinary. Mathematicians and physicists work together, to uncover the mysteries of this new discipline. In our experience it has seldom previously been the case that scientists have emerged from their own 'closed' realm of thought, and made their research results known to a broad lay public. That occurs typically here.

- These computer graphics, the results of mathematical research, are very surprising and have once more raised the question of what 'art' really is.

  Are these computer graphics to become a symbol of our 'hi-tech' age?

- For the first time in the history of science the distance between the utmost frontiers of research, and what can be understood by the 'man in the street', has become vanishingly small.

Normally the distance between mathematical research, and what is taught in schools, is almost infinitely large. But here the concerns of a part of today's mathematical research can be made transparent. That has not been possible for a long time.

Anyone can join in the main events of this new research area, and come to a basic understanding of mathematics. The central figure in the theory of dynamical systems, the *Mandelbrot set* - the so-called 'Gingerbread Man' - was discovered only in 1980. Today, virtually anyone who owns a computer can generate this computer graphic for themselves, and investigate how its hidden structures unravel.

1.1 Chaos and Dynamical Systems - What Are They?

An old farmer's saying runs like this: 'When the cock crows on the dungheap, the weather will either change, or stay as it is.' Everyone can be 100 per cent correct with this weather forecast. We obtain a success rate of 60 per cent if we use the rule that tomorrow's weather will be the same as today's. Despite satellite photos, worldwide measuring networks for weather data, and supercomputers, the success rate of computer-generated predictions stands no higher than 80 per cent.

Why is it not better?

Why does the computer – the very incarnation of exactitude – find its limitations here?

Let us take a look at how meteorologists, with the aid of computers, make their predictions. The assumptions of the meteorologist are based on the *causality principle*. This states that equal causes produce equal effects – which nobody would seriously doubt. Therefore the knowledge of all weather data must make an exact prediction possible. Of course this cannot be achieved in practice, because we cannot set up measuring stations for collecting weather data in an arbitrarily large number of places. For this reason the meteorologists appeal to the *strong causality principle*, which holds that similar causes produce similar effects. In recent decades theoretical models for the changes in weather have been derived from this assumption.

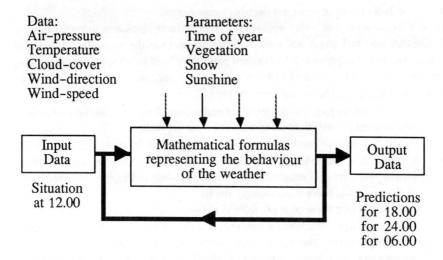

Figure 1.1-1 Feedback cycle of weather research.

Such models, in the form of complicated mathematical equations, are calculated with the aid of the computer and used for weather prediction. In practice weather data from the worldwide network of measuring stations, such as pressure, temperature, wind direction, and many other quantities, are entered into the computer system, which calculates the resulting weather with the aid of the underlying model. For example, in principle the method for predicting weather 6 hours ahead is illustrated in Figure 1.1-1. The 24-hour forecast can easily be obtained, by feeding the data for the 18-hour computation back into the model. In other words, the computer system generates output data with the aid of the weather forecasting program. The data thus obtained are fed back in again as input data. They produce new output data, which can again be treated as input data. The data are thus repeatedly fed back into the program.

One might imagine that the results thus obtained become ever more accurate. The opposite can often be the case. The computed weather forecast, which for several days has matched the weather very well, can on the following day lead to a catastrophically false prognosis. Even if the 'model system weather' gets into a 'harmonious' relation to the predictions, it can sometimes appear to behave 'chaotically'. The stability of the computed weather forecast is severely over-estimated, if the weather can change in unpredictable ways. For meteorologists, no more stability or order is detectable in such behaviour. The model system 'weather' breaks down in apparent disorder, in 'chaos'. This phenomenon of unpredictablity is characteristic of complex systems. In the transition from 'harmony' (predictability) into 'chaos' (unpredictability) is concealed the secret for understanding both concepts.

The concepts 'chaos' and 'chaos theory' are ambiguous. At the moment we agree to speak of chaos only when 'predictability breaks down'. As with the weather (whose correct prediction we classify as an 'ordered' result), we describe the meteorologists - often unfairly - as 'chaotic', when yet again they get it wrong.

Such concepts as 'order' and 'chaos' must remain unclear at the start of our investigation. To understand them we will soon carry out our own experiments. For this purpose we must clarify the many-sided concept of a *dynamical system.*

In general by a *system* we understand a collection of elements and their effects on each other. That seems rather abstract. But in fact we are surrounded by systems.

The weather, a wood, the global economy, a crowd of people in a football stadium, biological populations such as the totality of all fish in a pond, a nuclear power station: these are all systems, whose 'behaviour' can change very rapidly. The elements of the dynamical system 'football stadium', for example, are people: their relations with each other can be very different and of a multifaceted kind.

Real systems signal their presence through three factors:

- They are *dynamic*, that is, subject to lasting changes.
- They are *complex*, that is, depend on many parameters.
- They are *iterative*, that is, the laws that govern their behaviour can be described by feedback.

Today nobody can completely describe the interactions of such a system through mathematical formulas, nor predict the behaviour of people in a football stadium.

Despite this, scientists try to investigate the regularities that form the basis of such dynamical systems. In particular one exercise is to find simple mathematical models, with whose help one can simulate the behaviour of such a system.

We can represent this in schematic form as in Figure 1.1-2.

Of course in a system such as the weather, the transition from order to chaos is hard to predict. The cause of 'chaotic' behaviour is based on the fact that negligible changes to quantities that are coupled by feedback can produce unexpected chaotic effects. This is an apparently astonishing phenomenon, which scientists of many disciplines have studied with great excitement. It applies in particular to a range of problems that might bring into question recognised theories or stimulate new formulations, in biology, physics, chemistry and mathematics, and also in economic areas.

The research area of dynamical systems theory is manifestly interdisciplinary. The theory that causes this excitement is still quite young and – initially – so simple mathematically that anyone who has a computer system and can carry out elementary programming tasks can appreciate its startling results.

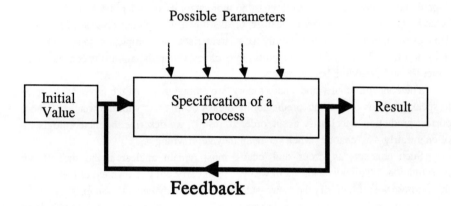

Figure 1.1–2 General feedback scheme.

The aim of chaos research is to understand in general how the transition from order to chaos takes place.

An important possibility for investigating the sensitivity of chaotic systems is to represent their behaviour by computer graphics. Above all, graphical representation of the results and independent experimentation has considerable aesthetic appeal, and is exciting.

In the following chapters we will introduce you to such experiments with different dynamical systems and their graphical representation. At the same time we will give you – a bit at a time – a vivid introduction to the conceptual world of this new research area.

1.2 Computer Graphics Experiments and Art

In their work, scientists distinguish two important phases. In the ideal case they alternate between experimental and theoretical phases. When scientists carry out an experiment, they pose a particular question to Nature. As a rule they offer a definite point of departure: this might be a chemical substance or a piece of technical apparatus, with which the experiment should be performed. They look for theoretical interpretations of the answers, which they mostly obtain by making measurements with their instruments.

For mathematicians, this procedure is relatively new. In their case the apparatus or measuring instrument is a computer. The questions are presented as formulas, representing a series of steps in an investigation. The results of measurement are numbers, which must be interpreted. To be able to grasp this multitude of numbers, they must be represented clearly. Often graphical methods are used to achieve this. Bar-charts and pie-charts, as well as coordinate systems with curves, are widespread examples. In most cases not only is a picture 'worth a thousand words': the picture is perhaps the only way to show the precise state of affairs.

Over the last few years experimental mathematics has become an exciting area, not just for professional researchers, but for the interested layman. With the availability of efficient personal computers, anyone can explore the new territory for himself.

The results of such computer graphics experiments are not just very attractive visually – in general they have never been produced by anyone else before.

In this book we will provide programs to make the different questions from this area of mathematics accessible. At first we will give the programs at full length; but later – following the building-block principle – we shall give only the new parts that have not occurred repeatedly.

Before we clarify the connection between experimental mathematics and computer graphics, we will show you some of these computer graphics. Soon you will be producing these, or similar, graphics for yourself. Whether they can be described as computer art you must decide for yourself.

Figure 1.2-1 Rough Diamond.

Figure 1.2-2   Vulcan's Eye.

Figure 1.2-3   Gingerbread Man.

Figure   1.2-4   Tornado Convention.[2]

Figure 1.2–5 Quadruple Alliance.

Figure 1.2-6 Seahorse Roundelay.

Figure 1.2–7   Julia Propeller.

Figure 1.2–8   Variation 1.

Figure 1.2–9   Variation 2.

Figure 1.2-10   Variation 3.

Figure 1.2-11   Explosion.

Figure   1.2–12   Mach 10.

Computer graphics in, computer art out. In the next chapter we will explain the relation between experimental mathematics and computer graphics. We will generate our own graphics and experiment for ourselves.

2 Between Order and Chaos: Feigenbaum Diagrams

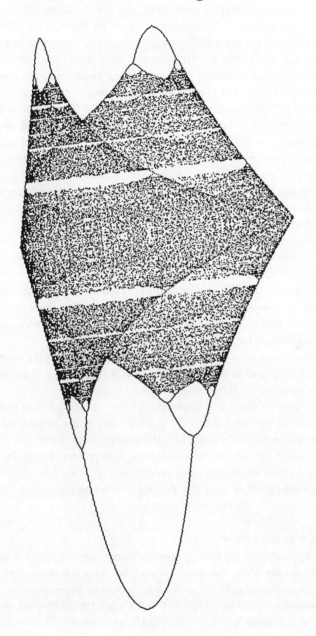

2.1   First Experiments

One of the most exciting experiments, in which we all take part, is one which Nature carries out upon us. This experiment is called *life*. The rules are the presumed laws of Nature, the materials are chemical compounds, and the results are extremely varied and surprising.   And something else is worth noting: if we view the ingredients and the product as equals, then each year (each day, each generation) begins with exactly what the previous year (day, generation) has left as the starting-point for the next stage. That development is possible in such circumstances is something we observe every day.

If we translate the above experiment into a mathematical one, then this is what we get: a fixed rule, which transforms input into output; that is, a rule for calculating the output by applying it to the input. The result is the input value for the second stage, whose result becomes the input for the third stage, and so on.   This mathematical principle of re-inserting a result into its own method of computation is called *feedback* (see Chapter 1).

We will show by a simple example that such feedback is not only easy to program, but it leads to surprising results. Like any good experiment, it raises ten times as many new questions as it answers.

The rules that will concern us are mathematical formulas. The values that we obtain will be real numbers between 0 and 1.   One possible meaning for numbers between 0 and 1 is as percentages: $0\% \le p \le 100\%$. Many of the rules that we describe in this book arise only from the mathematician's imagination. The rule described here originated when researchers tried to apply mathematical methods to growth, employing an interesting and widespread model. We will use the following as an example, taking care to remember that not everything in the model is completely realistic.

There has been an outbreak of measles in a children's home. Every day the number of sick children is recorded, because it is impossible to avoid sick and well children coming into contact with each other. *How does the number change?*

This problem corresponds to a typical dynamical system – naturally a very simple one. We will develop a mathematical model for it, which we can use to simulate an epidemic process, to understand the behaviour and regularities of such a system.

If, for example, 30% of the children are already sick, we represent this fact by the formula $p = 0.3$. The question arises, how many children will become ill the next day? The rule that describes the spread of disease is denoted mathematically by $f(p)$. The epidemic can then be described by the following equation:

$$f(p) = p + z.$$

That is, to the original p we add a growth z.

The value of z, the increase in the number of sick children, depends on the number p of children who are already sick. Mathematically we can write this dependence as $z \approx p$, saying that 'z is proportional to p'.   By this proportionality expression we mean that z may depend upon other quantities than p. We can predict that z depends also upon the number of well children, because there can be no increase if all the children are already sick in bed. If 30% are ill, then there must be 100% - 30% = 70% who are well. In general there will be 100%-p = 1-p well children, so we also have $z \approx$ (1-p). We have thus decided that $z \approx p$ and $z \approx$ (1-p). Combining these, we get a growth term $z \approx p*(1-p)$. But because not all children meet each other, and not every contact leads to an infection, we should include in the formula an *infection rate k*. Putting all this together into a single formula we find that:

$$z = k*p*(1-p),$$

so that

$$f(p) = p+k*p*(1-p).$$

In our investigation we will apply this formula on many successive days. In order to distinguish the numbers for a given day, we will attach an index to p. The initial value is p_0, after one day we have p_1, and so on. The result $f(p)$ becomes the initial value for the next stage, so we get the following scheme:

$$f(p_0) = p_0+k*p_0*(1-p_0) = p_1$$
$$f(p_1) = p_1+k*p_1*(1-p_1) = p_2$$
$$f(p_2) = p_2+k*p_2*(1-p_2) = p_3$$
$$f(p_3) = p_3+k*p_3*(1-p_3) = p_4$$

and so on. In general we have

$$f(p_n) = p_n+k*p_n*(1-p_n) = p_{n+1}.$$

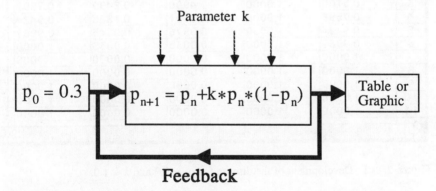

Figure 2.1-1 Feedback scheme for 'Measles'.

In other words this means nothing more than that the new values are computed from the old ones by applying the given rule. This process is called *mathematical feedback* or *iteration*. We have already spoken of this iterative procedure in our fundamental considerations in Chapter 1.

For any particular fixed value of k we can calculate the development of the disease from a given starting value p_0. Using a pocket calculator, or mental arithmetic, we find that these function values more or less quickly approach the limit 1; that is, all children fall sick. We would naturally expect this to occur faster, the larger the factor k is.

	Table	2-1		
	A	B	C	D	E
	p n	k	1 - p n	k*pn*(1-pn)	pn+1
1					
2	0.3000	0.5000	0.7000	0.1050	0.4050
3	0.4050	0.5000	0.5950	0.1205	0.5255
4	0.5255	0.5000	0.4745	0.1247	0.6502
5	0.6502	0.5000	0.3498	0.1137	0.7639
6	0.7639	0.5000	0.2361	0.0902	0.8541
7	0.8541	0.5000	0.1459	0.0623	0.9164
8	0.9164	0.5000	0.0836	0.0383	0.9547
9	0.9547	0.5000	0.0453	0.0216	0.9763
10	0.9763	0.5000	0.0237	0.0116	0.9879
11	0.9879	0.5000	0.0121	0.0060	0.9939

Figure 2.1-2 Development of the disease for $p_0 = 0.3$ and $k = 0.5$.

	Table	2-2		
	A	B	C	D	E
	p n	k	1 - p n	k*pn*(1-pn)	pn+1
1					
2	0.3000	1.0000	0.7000	0.2100	0.5100
3	0.5100	1.0000	0.4900	0.2499	0.7599
4	0.7599	1.0000	0.2401	0.1825	0.9424
5	0.9424	1.0000	0.0576	0.0543	0.9967
6	0.9967	1.0000	0.0033	0.0033	1.0000
7	1.0000	1.0000	0.0000	0.0000	1.0000
8	1.0000	1.0000	0.0000	0.0000	1.0000
9	1.0000	1.0000	0.0000	0.0000	1.0000
10	1.0000	1.0000	0.0000	0.0000	1.0000
11	1.0000	1.0000	0.0000	0.0000	1.0000

Figure 2.1-3 Development of the disease for $p_0 = 0.3$ and $k = 1.0$.

In order to get a feeling for the method of calculation, get out your pocket calculator. Work out the results first yourself, for the k-values

$$k_1 = 0.5, k_2 = 1, k_3 = 1.5, k_4 = 2, k_5 = 2.5, k_6 = 3$$

using the formula

$$f(p_n) = p_n + k*p_n*(1-p_n) = p_{n+1}$$

to work out p_1 up to p_5. Take $p_0 = 0.3$ in each case. So that you can check your results, we have given the calculation in the form of six tables (see Figures 2.1-2 to 2.1-7). In each table ten values per column are shown. In column A are the values p_i, in column E the values p_{i+1}.

Table 2-3					
	A	B	C	D	E
1	p n	k	1 - p n	k*pn*(1-pn)	pn+1
2	0.3000	1.5000	0.7000	0.3150	0.6150
3	0.6150	1.5000	0.3850	0.3552	0.9702
4	0.9702	1.5000	0.0298	0.0434	1.0136
5	1.0136	1.5000	-0.0136	-0.0207	0.9929
6	0.9929	1.5000	0.0071	0.0105	1.0035
7	1.0035	1.5000	-0.0035	-0.0052	0.9983
8	0.9983	1.5000	0.0017	0.0026	1.0009
9	1.0009	1.5000	-0.0009	-0.0013	0.9996
10	0.9996	1.5000	0.0004	0.0007	1.0002
11	1.0002	1.5000	-0.0002	-0.0003	0.9999

Figure 2.1-4 Development of the disease for $p_0 = 0.3$ and $k = 1.5$.

Table 2-4					
	A	B	C	D	E
1	p n	k	1 - p n	k*pn*(1-pn)	pn+1
2	0.3000	2.0000	0.7000	0.4200	0.7200
3	0.7200	2.0000	0.2800	0.4032	1.1232
4	1.1232	2.0000	-0.1232	-0.2768	0.8464
5	0.8464	2.0000	0.1536	0.2600	1.1064
6	1.1064	2.0000	-0.1064	-0.2354	0.8710
7	0.8710	2.0000	0.1290	0.2248	1.0957
8	1.0957	2.0000	-0.0957	-0.2098	0.8859
9	0.8859	2.0000	0.1141	0.2021	1.0880
10	1.0880	2.0000	-0.0880	-0.1916	0.8965
11	0.8965	2.0000	0.1035	0.1857	1.0821

Figure 2.1-5 Development of the disease for $p_0 = 0.3$ and $k = 2.0$.

Table 2-5					
	A	B	C	D	E
1	p n	k	1 - p n	k*pn*(1-pn)	pn+1
2	0.3000	2.5000	0.7000	0.5250	0.8250
3	0.8250	2.5000	0.1750	0.3609	1.1859
4	1.1859	2.5000	-0.1859	-0.5513	0.6347
5	0.6347	2.5000	0.3653	0.5797	1.2143
6	1.2143	2.5000	-0.2143	-0.6507	0.5637
7	0.5637	2.5000	0.4363	0.6149	1.1785
8	1.1785	2.5000	-0.1785	-0.5260	0.6525
9	0.6525	2.5000	0.3475	0.5669	1.2194
10	1.2194	2.5000	-0.2194	-0.6687	0.5507
11	0.5507	2.5000	0.4493	0.6186	1.1692

Figure 2.1-6 Development of the disease for $p_0 = 0.3$ and $k = 2.5$.

Table 2-6

	A	B	C	D	E
1	p n	k	1 - p n	k*pn*(1-pn)	pn+1
2	0.3000	3.0000	0.7000	0.6300	0.9300
3	0.9300	3.0000	0.0700	0.1953	1.1253
4	1.1253	3.0000	-0.1253	-0.4230	0.7023
5	0.7023	3.0000	0.2977	0.6272	1.3295
6	1.3295	3.0000	-0.3295	-1.3143	0.0152
7	0.0152	3.0000	0.9848	0.0449	0.0601
8	0.0601	3.0000	0.9399	0.1694	0.2295
9	0.2295	3.0000	0.7705	0.5305	0.7600
10	0.7600	3.0000	0.2400	0.5473	1.3072
11	1.3072	3.0000	-0.3072	-1.2048	0.1024

Figure 2.1-7   Development of the disease for p_0 = 0.3 and k = 3.0.

The tables were computed using the spreadsheet program 'Excel' on a Macintosh. Other spreadsheets, for example 'Multiplan', can also be used for this kind of investigation. For those interested, the program is given in Figure 2.1-8, together with the linking formulas. All diagrams involve the mathematical feedback process. The result of field E2 provides the starting value for A3, the result of E3 is the initial value for A4, and so on.

Measles Spreadsheet

	A	B	C	D	E
1	p n	k	1 - p n	k*pn*(1-pn)	pn+1
2	0.3	3	=1-A2	=B2*A2*C2	=A2+D2
3	=E2	=B2	=1-A3	=B3*A3*C3	=A3+D3
4	=E3	=B3	=1-A4	=B4*A4*C4	=A4+D4
5	=E4	=B4	=1-A5	=B5*A5*C5	=A5+D5
6	=E5	=B5	=1-A6	=B6*A6*C6	=A6+D6
7	=E6	=B6	=1-A7	=B7*A7*C7	=A7+D7
8	=E7	=B7	=1-A8	=B8*A8*C8	=A8+D8
9	=E8	=B8	=1-A9	=B9*A9*C9	=A9+D9
10	=E9	=B9	=1-A10	=B10*A10*C10	=A10+D10
11	=E10	=B10	=1-A11	=B11*A11*C11	=A11+D11

Figure 2.1-8   List of formulas.

Now represent your calculations graphically. You have six individual calculations to deal with. Each diagram, in a suitable coordinate system, contains a number of points generated by feedback.

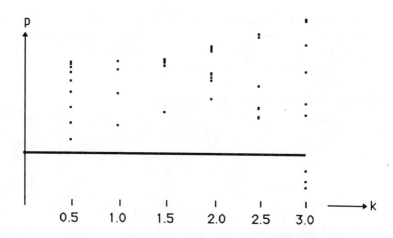

Figure 2.1-9 Discrete series of 6 (k_i, p_i)-values after 10 iterations.

We can combine all six diagrams into one, where for each k_i-value (k_i = 0.5, 1.0, 1.5, 2.0, 2.5, 3.0) we show the corresponding p_i-values (Figure 2.1-9).

You must have noticed how laborious all this is. Further, very little can be deduced from this picture. To gain an understanding of this dynamical system, it is not sufficient to carry out the feedback process for just 6 k-values. We must do more: for each k_i-value $0 \le k_i \le 3$ that can be distinguished in the picture, we must run continuously through the entire range of the k-axis, and draw in the corresponding p-values.

That is a tolerably heavy computation. No wonder that it took until the middle of this century before even such simple formulas were studied, with the help of newfangled computers.

A computer will also help us investigate the 'measles problem'. It carries out the same tedious, stupid calculation over and over again, always using the same formula.

When we go on to write a program in Pascal, it will be useful for more than just this problem. We construct it so that we can use large parts of it in other problems. New programs will be developed from this one, in which parts are inserted or removed. We just have to make sure that they fit together properly (see Chapter 11).

For this problem we have developed a Pascal program, in which only the main part of the problem is solved. Any of you who cannot finish the present problem, given the program, will find a complete solution in Chapters 11ff.

Program 2.1-1

```
PROGRAM MeaslesNumber;
    VAR
        Population, Feedback : real;
```

```
      MaximalIteration : integer;
(*-------------------------------------------------------*)
(*    BEGIN : Problem-specific procedures *)
      FUNCTION f (p, k : real) : real;
      BEGIN
         f := p + k * p * (1 - p) ;
      END;


      PROCEDURE MeaslesValue;
         VAR
            i : integer;
      BEGIN
         FOR i := 1 to MaximalIteration DO
         BEGIN
            Population := f(Population, Feedback);
            writeln('After' , i , 'Iterations p has the
                         value :',
            Population : 6 : 4);
         END;
      END;
   (* END Problem-specific procedures *)
 (* -------------------------------------------------------*)
 (* BEGIN: Useful subroutines *)
 (* see Chapter 11.2 *)
 (* END   : Useful subroutines *)


 (* BEGIN : Procedures of main program *)
   PROCEDURE Hello;
   BEGIN
      InfoOutput ('Calculation of Measles-Values');
      InfoOutput ('--------------------------');
      Newlines (2);
      CarryOn ('Start : ');
      Newlines (2);
   END;


   PROCEDURE Application;
   BEGIN
      ReadReal ('Initial Population p (0 to 1) >',
         Population);
      ReadReal ('Feedback Parameter k (0 to 3) >',
```

```
            Feedback);
      ReadInteger ('Max. Iteration Number >',
         MaximalIteration);
   END;


   PROCEDURE ComputeAndDisplay;
   BEGIN
      MeaslesValues;
   END;


   PROCEDURE Goodbye;
   BEGIN
      CarryOn ('To stop : ');
   END;
(* END : Procedures of Main Program)


BEGIN (* Main Program *)
   Hello;
   Application;
   ComputeAndDisplay;
   Goodbye;
END.
```

We have here written out only the main part of the program. The 'useful subroutines' are particular procedures to read in numbers or to output text to the screen (see Chapters 11ff.).

When we type in this Pascal program and run it, it gives an output like Figure 2.1-10. In Figure 2.1-10 not all iterations are shown. In particular the interesting values are missing. You should now experiment yourself: we invite you to do so before reading on. Only in this way can you appreciate blow by blow the world of computer simulation.

We have now built our first measuring instrument, and we can use it to make systematic investigations. What we have previously accomplished with tedious computations on a pocket calculator, have listed in tables, and drawn graphically (Figure 2.1-9) can now be done much more easily. We can carry out the calculations on a computer. We recommend that you now go to your computer and do some experimenting with Pascal program 2.1-1.

A final word about our 'measuring instrument'. The basic structure of the program, the main program, will not be changed much. It is a kind of standard tool, which we always construct. The useful subroutines are like machine parts or building blocks, which we can use in future, without worrying further. For those of you who do not feel so sure of yourselves we have an additional offer: complete tested programs and parts of programs. These are systematically collected together in Chapter 11.

```
Initial Population p (0 to 1)        >0.5
Feedback Parameter k (0 to 3)        >2.3
Max. Iteration No.                   >20


After 1  Iterations p has the value : 1.0750
After 2  Iterations p has the value : 0.8896
After 3  Iterations p has the value : 1.1155
After 4  Iterations p has the value : 0.8191
After 5  Iterations p has the value : 1.1599
After 6  Iterations p has the value : 0.7334
After 7  Iterations p has the value : 1.1831
After 8  Iterations p has the value : 0.6848
After 9  Iterations p has the value : 1.1813
After 10 Iterations p has the value : 0.6888
After 11 Iterations p has the value : 1.1818
After 12 Iterations p has the value : 0.6876
After 13 Iterations p has the value : 1.1817
After 14 Iterations p has the value : 0.6880
After 15 Iterations p has the value : 1.1817
After 16 Iterations p has the value : 0.6879
After 17 Iterations p has the value : 1.1817
After 18 Iterations p has the value : 1.6879
After 19 Iterations p has the value : 1.1817
After 20 Iterations p has the value : 0.6879
```

Figure 2.1-10   Calculation of measles values.

Computer Graphics Experiments and Exercises for §2.1

Exercise 2.1.-1

Implement the measles formula using a spreadsheet program. Generate similar tables to those shown in Figures 2.1-1 to 2.1-7. Check your values against the tables.

Exercise 2.1-2

Implement Program 2.1-1 on your computer. Carry out 30 iterations with 6 data sets. For a fixed initial value $p_0 = 0.3$ let k take values from 0 to 3 in steps of 0.5.

Exercise 2.1-3

Now experiment with other initial values of p, vary k, etc.

Once you've got the program `MeaslesNumber` running, you have your first measuring instrument. Find out for which values of k and which initial values of p the resulting series of p-values is

(a) simple (convergence to $p = 1$),

(b) interesting, and

(c) dangerous.

We call a series 'dangerous' when the values get larger and larger – so the danger is that they exceed what the computer can handle. For many implementations of Pascal the following range of values is not dangerous: $10^{-37} < |x| < 10^{38}$ for numbers x of type `real`.

By the interesting range of k we mean the interval from $k = 1.8$ to $k = 3.0$. Above this range it is dangerous; below, it is boring.

Exercise 2.1-4

Now that we have delineated the boundaries of the k-regions, we can present the above results acoustically. To do this you must change the program a little.

Rewrite Program 2.1-1 so that the series of numerical values becomes audible as a series of musical tones.

Exercise 2.1-5

What do you observe as a result of your experiments?

2.1.1 It's Prettier with Graphics

It can definitely happen that for some k-values no regularity can be seen in the series of numbers produced: the p-values seem to be more or less disordered. The experiment of Exercise 2.1-4 yields a regular occurrence of similar tone sequences only for certain values of p and k. So we will now make the computer sketch the results of our experiments, because we cannot find our way about this 'numerical salad' in any other manner. To do that we must first solve the problem of relating a cartesian coordinate system with coordinates x,y or k,p to the screen coordinates. Consider Figure 2.1.1-1 below.

Our graphical representations must be transformed in such a way that they can all be drawn on the same screen. In the jargon of computer graphics we refer to our mathematical coordinate system as the *universal coordinate system*. With the aid of a transformation equation we can convert the universal coordinates into screen coordinates.

Figure 2.1.1-1 shows the general case, in which we wish to map a *window*, or rectangular section of the screen, onto a projection surface, representing part of the screen. The capital letter U represents the universal coordinate system, and S the screen

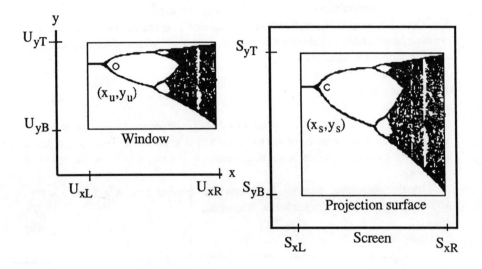

Figure 2.1.1-1 Two coordinate systems.

coordinate system.

The following transformation equations hold:

$$x_S = \frac{S_{xR}-S_{xL}}{U_{xR}-U_{xL}}(x_u-U_{xL})+S_{xL} \; ,$$

$$y_S = \frac{S_{yT}-S_{yB}}{U_{yT}-U_{yB}}(y_u-U_{yB})+S_{yB} \; .$$

Here L, R, B, T are the initials of 'left', 'right', 'bottom', 'top'. We want to express the transformation equation as simply as possible. To do this, we assume that we wish to map the window onto the entire screen. Then we can make the following definitions:

- U_{yT} = Top   and S_{yT} = Yscreen
- U_{yB} = Bottom   and S_{yB} = 0
- U_{xL} = Left   and S_{xL} = 0
- U_{xR} = Right   and S_{xR} = Xscreen.

This simplifies the transformation equation:

$$x_S = \frac{Xscreen}{Right-Left}(x_u-Left)$$

$$y_S = \frac{Yscreen}{Top-Bottom}(y_u-Bottom) \; .$$

On the basis of this formula we will write a program that is suitable for displaying the measles values. Observe its similar structure to that of Program 2.1-1.

Program   2.1.1-1

```
PROGRAM MeaslesGraphic;
   (* Possible declaration of graphics library *)
   (* Insert in a suitable place *)
CONST
   Xscreen = 320;   (* e.g. 320 pixels in x-direction *)
   Yscreen = 200;   (* e.g. 200 pixels in y-direction *)
VAR
   Left, Right, Top, Bottom, Feedback : real;
   IterationNo : Integer;
(* BEGIN:  Graphics Procedures *)
PROCEDURE SetPoint (xs, ys : integer);
BEGIN  (* Insert machine-specific graphics commands here*)
END;

PROCEDURE SetUniversalPoint (xu, yu: real);
   VAR
      xs, ys : real;
BEGIN
   xs := (xu - Left) * Xscreen / (Right - Left);
   ys := (yu - Bottom) * Yscreen / (Top - Bottom);
   SetPoint (round(xs), round(ys));
END;

PROCEDURE TextMode;
BEGIN
   (* Insert machine-specific commands: see hints *)
   (* in Chapter 11 *)
END;

PROCEDURE GraphicsMode;
BEGIN
   (* Insert machine-specific commands: see hints *) in
   (* Chapter 11 *)
END;
PROCEDURE EnterGraphics;
   (* various actions to initialise the graphics *)
   (* such as GraphicsMode etc.                        *)
      GraphicsMode;
```

```
END;
PROCEDURE ExitGraphics;
BEGIN
(* Actions to end the graphics, e.g. : *)
   REPEAT
      (* Button is a machine-specific procedure *)
   UNTIL Button;
   TextMode;
END;
(* END:  Graphics Procedures *)
(* ------------------------------------------------------ *)
(* BEGIN : Problem-specific Procedures *)
FUNCTION f (p, k : real) : real;
BEGIN
   f := p + k * p * (1 - p);
END;

PROCEDURE MeaslesIteration;
   VAR
      range, i: integer
      population : real
      deltaxPerPixel: real;
BEGIN
   deltaxPerPixel := (Right - Left) / Xscreen;
   FOR range := 0 TO Xscreen DO
   BEGIN
      Feedback := Left + range * deltaxPerPixel;
      population := 0.3
      FOR i := 0 to IterationNo DO
      BEGIN
         SetUniversalPoint (Feedback, population);
         population := f ( population, Feedback );
      END;
   END;
END;
(* END:  Problem-specific Procedures *)
(*-----------------------------------------------------*)
(* BEGIN  Useful Subroutines *)
(* See Program 2.1-1, not given here *)
(* END : Useful Subroutines *)
(* BEGIN: Procedures of Main Program *)
```

```
PROCEDURE Hello;
BEGIN
   TextMode;
   InfoOutput ('Diagram of the Measles Problem');
   InfoOutput ('-----------------------------');
   Newlines (2);
   CarryOn ('Start : ');
   Newlines (2);
END;

PROCEDURE Initialise;
BEGIN
   ReadReal ('Left                    >', Left);
   ReadReal ('Right                   >', Right);
   ReadReal ('Top                     >', Top);
   ReadReal ('Bottom                  >', Bottom);
   ReadInteger (Iteration Number      >', IterationNo);
END;

PROCEDURE ComputeAndDisplay;
BEGIN
   EnterGraphics;
   MeaslesIteration;
   ExitGraphics;
   END;

PROCEDURE Goodbye;
BEGIN
   CarryOn ('To end : ');
END
(*END : Procedures of Main Program *)

BEGIN (* Main Program *)
   Hello;
   Initialise;
   ComputeAndDisplay;
   Goodbye;
END.
```

We suggest that you now formulate Program 2.1.1-1 as a complete Pascal program and enter it into your machine. The description above may help, but you may have developed your own programming style, in which case you can do everything differently if you

wish.   Basically Program 2.1.1-1 solves the 'algorithmic heart' of the problem. The
machine–specific components are discussed in Chapter 12 in the form of reference
programs with the appropriate graphics commands included.

TextMode, GraphicsMode, and Button are machine-specific procedures. In
implementations, TextMode and GraphicsMode are system procedures. This is the
case for Turbo Pascal on MS–DOS machines and for UCSD Pascal (see Chapter 12).

Button corresponds to the Keypressed function of Turbo Pascal. The 'useful
subroutines' have already been described in Program 2.2-1.   By comparing Programs
2.1-1 and 2.1.1-1 you will see that we have converted our original 'numerical'
measuring–instrument into a 'graphical' one.   Now we can visualise the number flow
more easily.

The development of the program mostly concerns the graphics: the basic structure
remains unchanged.

Something   new,   which   we   must   clarify,   occurs   in   the   procedure
MeaslesIteration (see Program 2.1.1-1):

```
deltaxPerPixel := (Right - Left) / Xscreen;
   FOR range := 0 TO Xscreen DO
   BEGIN
      Feedback := Left + range * deltaxPerPixel;
                . . .
```

Compare this with the transformation formula:

$$x_s = \frac{Xscreen}{Right-Left}(x_u-Left)$$

Solve this equation for x_s.

When we give the screen coordinate x_s the value 0, then the universal coordinate
must become Left.   Setting the value Xscreen for the maximal screen coordinate x_s,
we get the value Right.   Every other screen coordinate corresponds to a universal
coordinate between Left and Right. The smallest unit of size that can be distinguished
on the screen is one pixel.   The corresponding smallest unit of size in universal
coordinates is thus deltaxPerPixel.

After this brief explanation of the graphical representations of the measles data with
the aid of Program 2.1.1-1, we will describe the result, produced by the computer
program in the form of a graphic. See Figure 2.1.1-2, to which we have added the
coordinate axes.

How do we interpret this graphic?   From right to left the factor k changes in the
range 0 to 3.   For small values of k (in particular $k = 0$) the value of p changes by little
or nothing. For k-values in the region of 1 we see the expected result: p takes the value
1 and no further changes occur.

The interpretation for the model is thus: if the infection rate k is sufficiently large,
then soon all children become ill ($p = 100\%$). This occurs more rapidly, the larger k is.

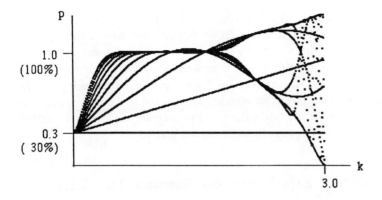

Figure 2.1.1-2 Representation of the measles epidemic on the screen,
IterationNo = 10.

You can also see this result using the values computed by pocket calculator, e.g. Figures 2.1-1 to 2.1-7.

For k-values greater than 1 something surprising and unexpected happens: p can become larger than 1! Mathematically, that's still meaningful. Using the formula you can check that the calculation has proceeded correctly. Unfortunately it illustrates a restriction on our measles example, because more than 100% of the children cannot become ill. The picture shows quite different results here. Might something abnormal be going on?

Here we find ourselves in a typical experimental situation: the experiment has to some extent confirmed our expectations, but has also led to unexpected results. That suggests new questions, which possess their own momentum. Even though we can't make sense of the statement that 'more than 100% of children become sick', the following question starts to look interesting: how does p behave, if k gets bigger than 2?

Figure 2.1.1-2 provides a hint: p certainly does not, as previously, tend to the constant value $p = 1$. Apparently there is no fixed value which p approaches, or, as mathematicians say, towards which the sequence p *converges*.

It is also worth noting that the sequence does not *diverge* either. Then p would increase beyond all bounds and tend towards +∞ or -∞. In fact the values of p jump about 'chaotically', to and fro, in a range of p between $p = 0$ and $p = 1.5$. It does not seem to settle down to any particular value, as we might have expected, but to many. What does that mean?

In order better to understand the number sequences for the population p, we will now take a quick look at the screen print-out of Figure 2.1-10 (calculation of measles values) from Chapter 2.1.

We can use the program again in Exercises 2.1-1 to 2.1-4, which we have already given, letting us display the results once more on the screen (see Figure 2.1-10). As an additional luxury we can also make the results audible as a series of musical tones. The melody is not important. You can easily tell whether the curve tends towards a single value or many. If we experiment on the `MeaslesNumber` program with $k = 2.3$, we find an ' oscillating phase' jumping to and fro between two values (see Figure 1.2–10). One value is > 1, and the other < 1. For $k = 2.5$ it is even more interesting. At this point you should stop hiding behind the skirts of our book, and we therefore suggest that, if you have not done so already, you write your first program and carry out your first experiment now. We will once more formulate the task precisely:

Computer Graphics Experiments and Exercises for §2.1.1

Exercise 2.1.1–1
Derive the general transformation equations for yourself with the aid of Figure 2.1.1–1. Check that the simplified equation follows from the general one. Explain the relation between them.

Exercise 2.1.1–2
Implement the Pascal program `MeaslesGraphic` on your computer. Check that you obtain the same graphic displays as in Figure 2.1.1–2. That shows you are on the right track.

Exercise 2.1.1.3
Establish the connection between the special transformation formula and the expression for `deltaxPerPixel`.

Exercise 2.1.1–4
Use the program `MeaslesGraphic` to carry out the same investigations as in Exercise 2.1–3 (see Chapter 2.1) – this time with graphical representation of the results.

2.1.2 Graphical Iteration
It may perhaps have occurred to you that the function
$$f(x) = x + k * x * (1-x)$$
– for so we can also write the equation – is nothing other than the function for a parabola
$$f(x) = -k * x^2 + (k+1) * x$$
This is the equation of a downward–opening parabola through the origin with its vertex in the first quadrant. It is clear that for different values of k we get different parabolas. We can also study the 'feedback effect' of this parabola equation by *graphical iteration*.

Let us explain this important concept.

Feedback means that the result of a calculation is replaced into the same equation as

a new initial condition.   After many such feedbacks (iterations) we establish that the
results run through certain fixed values. By *graphical feedback* we refer to the picture of
the function in an x,y-coordinate system (x stands for p, y for $f(x)$ or $f(p)$).

Graphical iteration takes place in the following way. Beginning with an x-value,
move vertically up or down in the picture until you hit the parabola. You can read off
$f(x)$ on the y-axis. This is the initial value for the next stage of the feedback. The
value must be carried across to the x-axis. For this purpose we use the diagonal, with
equation $y = x$. From the point on the parabola (with coordinates $(x, f(x))$) we draw a
line horizontally to the right (or left), until we encounter the diagonal (at the coordinates
$(f(x), f(x))$).   Then we draw another vertical to meet the parabola, a horizontal to meet
the diagonal, and so on.

This procedure will be explained further in the program and the pictures that follow
it.

Program Fragment 2.1.2–1

```
(* ------------------------------------------------- *)
(* BEGIN : Problem-specific Procedures *)
FUNCTION f (p, k : real) : real;
BEGIN
   f := p + k * p * (1 - p);
END;


PROCEDURE ParabolaAndDiagonal(population,feedback : real) ;
   VAR
       xCoord, deltaxPerPixel : real;
BEGIN
   DeltaxPerPixel : = (Right - Left) / Xscreen;
   SetUniversalPoint (Left, Bottom);
   DrawUniversalLine  (Right, Top);
   DrawUniversalLine (Left, Bottom);
   xCoord  := Left;
   REPEAT
      DrawUniversalLine (xCoord, f(xCoord, feedback));
      xCoord := xCoord + deltaxPerPixel;
   UNTIL (xCoord > Right);
   GoToUniversalPoint (population, Bottom);
END;


PROCEDURE GraphicalIteration;
(* Version for graphical iteration *)
   VAR
```

```
          previousPopulation : real;
     BEGIN
        ParabolaAndDiagonal (population, feedback);
        REPEAT
           DrawUniversalLine (population, population);
           previousPopulation := population;
           population := f(population, feedback);
           DrawUniversalLine (previousPopulation, population);
        UNTIL Button;
     END;

     (* END : Problem-specific Procedures *)
     (* ------------------------------------------------ *)


     (* DrawUniversalLine (x,y) draws a line from the *)
     (* current position to the point with universal coordinates
 (x,y). *)
```

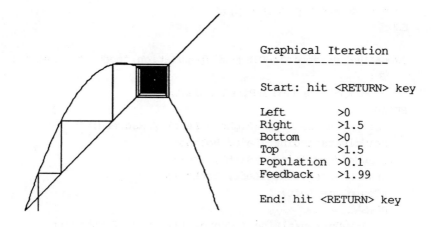

```
Graphical Iteration
-------------------

Start: hit <RETURN> key

Left          >0
Right         >1.5
Bottom        >0
Top           >1.5
Population    >0.1
Feedback      >1.99

End: hit <RETURN> key
```

Figure 2.1.2-1   Initial value $p = 0.1$, $k = 1.99$, a limit point, with screen dialogue.

For each given k-value we get distinct pictures. If the final value is the point $f(p) = 1$, we obtain a spiral track (Figure 2.1.2-1). In all other cases the horizontal and vertical lines tend towards segments of the original curve, which correspond to limiting p-values. Clearly the two vertical lines in Figure 2.1.2-2 represent two different p-values.

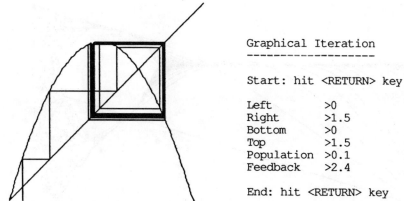

```
Graphical Iteration
-------------------

Start: hit <RETURN> key

Left          >0
Right         >1.5
Bottom        >0
Top           >1.5
Population    >0.1
Feedback      >2.4

End: hit <RETURN> key
```

Figure 2.1.2-2   Initial value $p = 0.1$, $k = 2.4$, two limiting points.

The distinct cases (limiting value 1, or n–fold cycles, Figures 2.1.2–1, 2.1.2–2) are thus made manifest. For an overview it can be useful to carry out the first 50 iterations without drawing them, after which 50 iterations are carried out and drawn.

This process of graphical iteration can also be applied to other functions. In this way we obtain rules, about the form of the graph of a function, telling us which of the above two effects it will produce.

Computer Graphics Experiments and Exercises for §2.1.2

Exercise 2.1.2–1

Develop a program for graphical iteration. Try to generate Figures 2.1.2–1 and 2.1.2–2. Experiment with the initial value $p = 0.1$ and $k = 2.5$, 3.0. How many limiting values do you get?

Exercise 2.1.2–2

Devise some other functions and apply graphical iteration to them.

2.2   Fig–trees Forever

In our experiments with the program `MeaslesGraphic` you must surely have noticed that the lines in the range $0 \le k \le 1$ get closer and closer together, if we increase the number of iterations (see Program 2.1.1–1). Until now we have computed with small values, in order not to occupy too much of the computer's time. But now we will make our first survey of the entire range. Figure 2.2.–1 shows the result of 50 iterations for comparison with Figure 2.1.1-2.

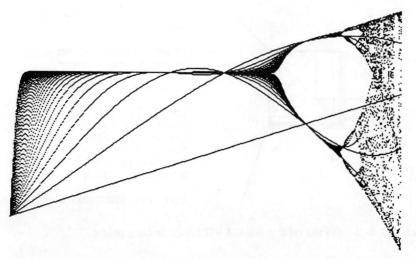

Figure 2.2-1 Situation after the onset of oscillations, iteration number = 50.

Obviously some structure comes to light when we increase the accuracy of our measurements (that is, the number of iterations). And it is also clear that the extra lines in the range $0 \leq k \leq 1$ are *transient effects*. If we first carry out the iteration procedure for a while (say 50 iterations) without drawing points, and then continue to iterate while plotting the resulting points, the lines will disappear.

The above remarks are in complete agreement with our fundamental ideas in the simulation of dynamical systems. We are interested in the 'long-term behaviour' of a system under feedback (see Chapter 1). Program 2.2.-1 shows how easily we can modify our program MeaslesGraphic, in order to represent the long-term behaviour more clearly.

Program Fragment 2.2-1

```
(* BEGIN:   Problem-specific procedures *)
FUNCTION f (p, k : real) : real;
BEGIN
    f := p + k * p * (1 - p);
END;
PROCEDURE FeigenbaumIteration;
    VAR
        range, i :integer;
        population, deltaxPerPixel : real;
BEGIN
    deltaxPerPixel := (Right - Left) / Xscreen;
    FOR range := 0 TO Xscreen DO
```

```
      BEGIN
          Feedback := Left + range*deltaxPerPixel;
          population := 0.3;
          FOR i := 0 to Invisible DO
             BEGIN
                 population := f(population, Feedback);
             END
          FOR i := 0 TO Visible DO
             BEGIN
                 SetUniversalPoint (Feedback, population);
                 population := f(population, Feedback);
             END;
      END;
   END;
   (* END: Problem-specific procedures *)
   (* --------------------------------------------------------*)
   PROCEDURE Initialise;
   BEGIN
      ReadReal ('Left          >', Left);
      ReadReal ('Right         >', Right);
      ReadReal ('Bottom        >', Bottom);
      ReadReal ('Top           >', Top);
      ReadInteger ('Invisible      >', Invisible);
      ReadInteger ('Visible        >'), Visible);
   END;

   PROCEDURE ComputeAndDisplay;
   BEGIN
      EnterGraphics;
      FeigenbaumIteration;
      ExitGraphics;
   END;
```

The new or modified parts of the program are shown in bold type. If we type in and run this progam then it gives a print-out as in Figure 2.2.-2. It shows for the 'interesting' range $k > 1.5$ a piece of the so-called *Feigenbaum diagram*.[1]

This program and picture will keep us busy for a while.

---

[1] *Translator's note*: This is more commonly known as a *bifurcation diagram*.

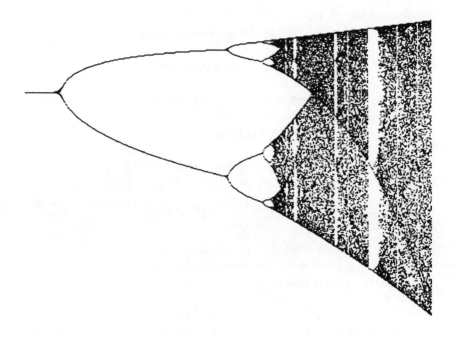

Figure 2.2-2   Print-out from Program 2.2-1.

- The name 'Feigenbaum' is in honour of the physicist Mitchell Feigenbaum, [2] who carried out the pioneering research described in this chapter. We shall call any picture like Figure 2.2-1 a *Feigenbaum diagram.*

- In the program fragment we introduce two new variables `Invisible` and `Visible`, which in the example are given the value 50.

The results show a certain independence of the initial value for p, provided we do not start with $p = 0$ or $p = 1$. You will probably have discovered that already. What interests us here is just the results of a large number of iterations. To stop the picture looking unsightly, the first 50 iterations run 'in the dark'. That is, we do not plot the results k,p. After that, a further 50 (or 100 or 200) iterations are made visible.

In order to facilitate comparison with Figure 2.2-1, you can set the variables in Program 2.2-1 as follows:

```
Invisible := 0; Visible := 10;
```

As regards the working of the program, the following remarks should be made:

Input data are read from the keyboard and assigned to the corresponding variables in the procedure `Initialise`. It is then easy to set up arbitrary values from the keyboard. However, the program must then be initialised on each run. The type of input

---

[2] *Translator's note*: It is also German for ' fig-tree' , hence the section title.

procedure used depends on the purpose of the program.   With keyboard input, typing
errors are possible. Sometimes it is useful to insert fixed values into the program.

To draw Figure 2.2-2 on the screen on an 8-bit home computer takes about 5-10
minutes.   With more efficient machines (Macintosh, IBM, VAX, SUN, etc.) it is
quicker.

It is harder to describe these astonishing pictures than it is to produce them slowly
on the screen.   What for small k converges so regularly to the number 1, cannot
continue to do so for larger values of k because of the increased growth-rate.   The
curve splits into two branches, then 4, then 8, 16, and so on.   We have discovered this
effect of 2, 4, or more branches (limiting values) by graphical iteration.   This
phenomenon is called a *period-doubling cascade*, (Peitgen and Richter 1986, p.7).

When $k > 2.570$ we see behaviour that can only be described by a new concept:
chaos. There are unpredictable 'flashes' of points on the screen, and no visible
regularity.

As we develop our investigations we will show that we have not blundered into
chaos by accident. We have witnessed the crossing of a frontier. Up to the point $k = 2$
our mathematical world is still ordered.   But if we work with the same formula and
without rounding errors, for higher values of k it is virtually impossible to predict the
outcome of the computation.   A series of iterations beginning with the value $p = 0.1$,
and one beginning with $p = 0.11$, can after a few iterations become completely
independent, exhibiting totally different behaviour. A small change in the initial state can
have unexpected consequences. 'Small cause, large effect': this statement moreover holds
in a noticeably large region. For our Feigenbaum formula the value $k = 2.57$ divides
'order and chaos' from each other. On the right-hand-side of Figure 2.2-2 there is no
order to be found. But this chaos is rather interesting – it contains structure!

Figure 2.2-2 appears to have been drawn by accident.   As an example, let us
consider the neighbourhood of the k-value 2.84. Here there is a region in which points
are very densely packed. On the other hand, there are also places nearby with hardly any
points at all.   By looking carefully we can discover interesting structures, in which
branching again plays a role.

In order to search for finer detail, we must 'magnify' the picture. On a computer
this means that we want to display a *window*, or *section*, from the full picture 2.2-2 on
the screen.[3] To do this we give suitable values to the variables `Right`, `Left`, `Bottom`,
and `Top`. The program user can input values from the keyboard.   In that way it is
possible to change the window at will, to investigate interesting regions. If the picture is
expanded a large amount in the y-direction it becomes very 'thin' , because the majority
of points lead outside the window.   It then makes sense, by changing the variable
`Visible`, to increase the total number of points plotted.

We now investigate the precise construction of the Feigenbaum diagram, with the
aid of a new program. It is derived by a small modification of Program Fragment 2.2-1.

---

[3]In the choice of a window there is often a problem, to find out the values for the edges. As a
simple aid we construct a transparent grid, which divides the screen into ten parts in each direction.

Program Fragment 2.2-2

```
. . . . .
deltaxPerPixel := (Right - Left) / Xscreen;
FOR range := 0 to Xscreen DO
   BEGIN
      Feedback := Left + range * deltaxPerPixel;
      DisplayFeedback (Feedback);
      population := 0.3;
. . . . .
```

Elsewhere we will introduce a procedure DisplayFeedback, and thereby enlarge our experimental possibilities. DisplayFeedback displays the current value of k in the lower left corner of the screen. It will be useful later, to establish more accurately the boundaries of interesting regions in the Feigenbaum diagram. To display text on the graphics screen some computers (such as the Apple II) require a special procedure. Other computers have the ability to display numbers of type real directly on the graphics screen, or to display text and graphics simultaneously in different windows.

The procedure DisplayFeedback can also be omitted if it is not desired to display numerical values on the screen. In this case DisplayFeedback must be deleted from the initial part of the main program, and also in the procedure FeigenbaumIteration which calls it.

When the program runs correctly, you should use it to draw sections of the Feigenbaum diagram. By choosing the boundaries of the windows suitably you can plot pictures whose fine detail can scarcely be distinguished in the full diagram. A tiny part of the picture can already contain the form of the whole. This astonishing property of the Feigenbaum diagram, containing itself, is called *self-similarity*. Look for yourself for further examples of self-similarity in the Feigenbaum diagram.

We should describe how the above program works in practice. Instructions appear on the screen for the input of the necessary data. The data are always input by using the <RETURN> key. The dialogue might, for example, go like this:

```
Start:       <hit RETURN key
Left         (>= 1.8)    >2.5
Right        (<= 3)      >2.8
Bottom       (>= 0)      >0.9
Top          (<= 1.5)    >1.4
Invisible    (>= 50)     >50
Visible      (>= 50)     >50
```

The picture that arises from this choice of input data is shown in Figure 2.2-3.

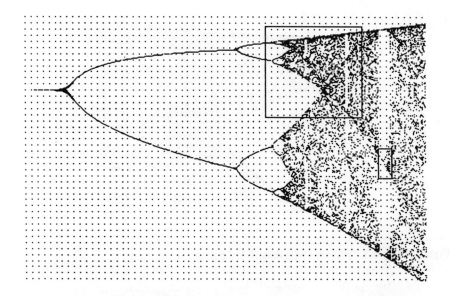

Figure   2.2-3 Section from the Feigenbaum diagram (see the following figures).

Figures 2.2-4 and 2.2-5 represent such sections from the Feigenbaum diagram, as drawn in Figure 2.2-3.

We also suggest that you try equations other than the Feigenbaum equation. Surprisingly, you will find that quite similar pictures appear! In many cases we find that the picture again begins with a line, and splits into 2, 4, 8,... twigs. There is also another common feature, which we do not wish to discuss further at this stage.

The stated values in Figures 2.2-4 and 2.2-5 are just examples of possible inputs. Try to find other interesting places for yourself.

Computer Graphics Experiments and Exercises for §2.2

Exercise   2.2-1

Implement Program 2.2-1 on your computer. Experiment with different values of the variables `Visible` and `Invisible`.

Exercise   2.2-2

Extend the program to include a procedure `DisplayFeedback`, which during the running of the program can 'measure' the k-values.

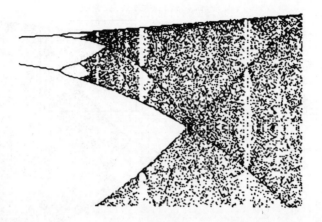

Figure   2.2-4   Fig-tree with data: 2.5, 2.8, 0.9, 1.4, 50, 100.

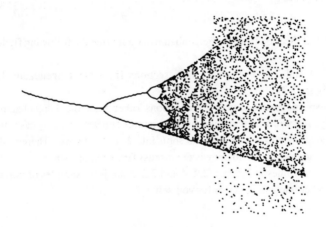

Figure   2.2-5   Fig-tree with data: 2.83, 2.87, 0.5, 0.8, 50, 100.

Exercise   2.2-3

Find regions in the Feigenbaum diagram around $k = 2.8$, where self-similarity can be found.

Exercise   2.2-4

Try to discover 'hidden structure', when you increase the iteration number in interesting regions. Think about taking small regions (and magnifying them).

Exercise   2.2-5

As regards the chaotic phenomena of Feigenbaum iteration, much more can be said.

'The initial value leads to unpredictable behaviour, but on average there is a definite result.' Test this hypothesis by displaying the average value of a large number of results as a graph, for k-values in the chaotic region.

See if you can confirm this hypothesis, or perhaps the contrary: 'Chaos is so fundamental that even the averages for k-values taken close together get spread out.'

Exercises   2.2-6

That after these explanations our investigation of the 'fig-tree' has not revealed all its secrets, is shown by the following consideration:

Why must the result of the function f always depend only on the previous p–value?

It is possible to imagine that the progenitors of this value 'have a say in the matter'. The value f_n for the nth iteration would then depend not only on f_{n-1}, but also on f_{n-2}, etc.   It would be sensible if 'older generations' had somewhat less effect.   In a program you can, for example, store the most recent value as pn, the previous one as pnMinus1, and the one before that as pnMinus2. The function f can then be viewed as follows. We give two examples in Pascal notation.

```
f(pn)    := pn + 1/2*k*(3*pn*(1-pn)-
                 pnMinus1*(1-pnMinus1));
```
or
```
f(pn)    := pn +1/2*k*(3*pnMinus1-pnMinus2)*
                 (1-3*pnMinus1-pnMinus2 );
```

To start, pn, pnMinus1, etc. should be given sensible values such as 0.3, and at each stage they should obviously be given their new values. The k-values must lie in a rather different range than previously. Try it out!

In the above print-out it goes without saying that other factors such as –1 and 3 and other summands are possible.   The equations under consideration no longer have anything to do with the original 'measles' problem. They are not entirely unknown to mathematicians: they appear in a similar form in approximation methods for the solution of differential equations.

Exercise   2.2-7

In summary we might say that we always obtain a Feigenbaum diagram if the recursion equation is *nonlinear*. In other words, the underlying graph must be curved.

The diagrams appear especially unusual, if more generations of values are made visible. This gives rise to a new set of functions to investigate, for which we can change the series, in which we 'worry about the important bend in the curve' - which happens to be the term expression* (1 - expression) into which we substitute the previous value:

```
f(pn) := pn + 1/2*k*(3*pnMinus1*(1-pnMinus1)-
         pnMinus2*(1-pnMinus2));
```

Exercise 2.2-8

Investigate at which k_i-values branches occur.

2.2.1   Bifurcation Scenario - the Magic Number 'Delta'

The splittings in the Feigenbaum diagram, which by now you have seen repeatedly in your experiments, are called *bifurcations*. In the Feigenbaum diagram illustrated above, some points, the *branch points*, play a special role. We use the notation k_i for these: k_1, k_2, and so on. We can read off from the figures that $k_1 = 2$, $k_2 = 2.45$, and $k_3 = 2.544$. You can obtain these results with some effort from Exercise 2.2-8.

It was the great discovery of Mitchell Feigenbaum to have found a connection between these numbers. He realised that the sequence

$$\frac{k_n - k_{n-1}}{k_{n+1} - k_n}, \text{ for } n = 2, 3, \dots$$

converges to a constant value δ (the Greek letter 'delta') when n tends to ∞. Its decimal expansion begins $\delta = 4.669 \dots$.

We have formulated a series of interesting exercises about this number δ (Exercises 2.2.1-1ff. at the end of this section). They are particularly recommended if you enjoy number games and are interested in 'magic numbers'. Incidentally, you will then have shown that δ is a genuinely significant mathematical constant, which appears in several contexts. This same number arises in many different processes involving dynamical systems. For bifurcation problems it is as characteristic as the number π is for the area

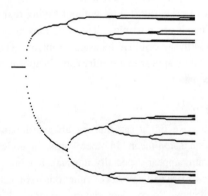

Figure 2.2.1-1 Logarithmic representation from $k = 1.6$ to $k = 2.569$.

and circumference of a circle. We call this number the *Feigenbaum number*. Mitchell Feigenbaum demonstrated its universality in many computer experiments. [4]

The higher symmetry and proportion that lies behind the above is especially significant if we do not choose a linear scale on the k-axis. Once the limiting value k_∞ of the sequence k_1, k_2, k_3, ... is known, a logarithmic scale is preferable.

Computer Graphics Experiments and Exercises for §2.2.1

Exercise 2.2.1-1

The Feigenbaum constant δ has proved to be a natural constant, which occurs in situations other than that in which Feigenbaum first discovered it. Compute this natural constant as accurately as possible:

$$\delta = \lim_{n \to \infty} \frac{k_n - k_{n-1}}{k_{n+1} - k_n} .$$

In order to work out δ, the values k_i must be calculated as accurately as possible. Using Program Fragments 2.2-1 and 2.2-2 you can look at the interesting intervals of k and p, and pursue the branching of the lines. By repeatedly magnifying windows taken from the diagram you can compute the k-values.

Near the branch-points, convergence is very bad. It can happen that even after 100 iterations we cannot decide whether branching has taken place.

We should henceforth make tiny changes to the program
- to make the point being worked on flash, and
- to avoid choosing a fixed iteration number at the start.

It is easy to make a point flash by changing its colour repeatedly from black to white and back again.

We can change the iteration number by using a different loop construction. Instead of

```
FOR counter := 1 to Visible DO
```
we introduce a construction of the form
```
REPEAT UNTIL Button; [5]
```

Exercise 2.2.1-2

Change the Feigenbaum program so that it uses a logarithmic scale for the k-axis instead of a linear one. Positions k should be replaced by $-\ln(k_\infty - k)$ measured from the right.

For the usual Feigenbaum diagram the limit k_∞ of the sequence k_1, k_2, k_3, ... has the value 2.570. If, for example, we divide each decade (interval between successive

---

[4]The universality was proved mathematically by Pierre Collet, Jean-Pierre Eckmann, and Oscar Lanford (1980).

[5]In Turbo Pascal you must use REPEAT UNTIL Keypressed;

powers of 10) into 60 parts, and draw three decades on the screen, there will be a figure 180 points wide.

If you also expand the scale in the vertical direction, you will have a good measuring instrument to develop the computation of the k_i-values.

Exercise 2.2.1-3

With a lot of patience you can set loose the 'order within chaos' - investigate the region around $k = 2.84$. Convince yourself that δ has the same value as in the range $k < 2.57$.

Exercise 2.2.1-4

Develop a program to search for the k_i-values automatically, which works not graphically, but numerically. Bear in mind that numerical calculations in Pascal rapidly run into limitations. The internal binary representation for a floating-point number uses 23 bits, which corresponds to about 6 or 7 decimal places.

This restriction clearly did not put Feigenbaum off - he evaluated the aforementioned constant δ as 4.669 201 660 910 299 097 ... .

On some computers it is possible to represent numbers more accurately. Look it up in the manual.

Exercise 2.2.1-5

Feigenbaum's amazing constant arises not only when we follow the branching from left to right (small k-values to large ones). The 'bands of chaos', which are densely filled with points, also split when we go from large k-values to small ones. A single connected band splits into 2, then 4, then 8, ... Compute the k-values where this occurs.

Show that the constant δ appears here too.

2.2.2 Attractors and Frontiers

The mathematical equation which lies at the basis of our first experiment was formulated by Verhulst as early as 1845. He studied the growth of a group of animals, for which a restricted living space is available. In this interpretation it becomes clear what a value $p > 1$ means. $p = 100\%$ means that every animal has the optimum living space available. More than 100% corresponds to overpopulation. The simple calculations we have performed for the measles problem already show how the population then develops. For normal values of k the population is cut back until the value 1 is reached. However, the behaviour is different if we start with negative or large numbers. Even after many steps the population no longer manages to reach 1.

Mathematicians, like other scientists, habitually develop new ideas in order to attack new and interesting phenomena. This takes us a little way into the imposing framework of technical jargon. With clearly defined concepts it is possible to describe clearly defined circumstances. We will now encounter one such concept.

In the absence of anything better, mathematicians have developed a concept to capture the behaviour of the numbers in the Feigenbaum scenario. The final value $p = 1$ is called an *attractor*, because it 'pulls the solutions of the equations' towards itself.

This can be clearly seen on the left-hand side of Figure 2.1.1-2. However many times we feed back the results p_n into the Feigenbaum equation, all the results tend towards the magic final value 1. The p-values are drawn towards the attractor 1. What you may perhaps have noticed already in your experiments is another attractor, $-\infty$ (minus infinity). At higher values ($k > 2$) of the feedback constant, the finite attractor is not just the value 1. Consider the picture of the Feigenbaum diagram: *the whole figure is the attractor!*

Each sequence of p-values which starts near the attractor invariably ends with a sequence of numbers that belong to the attractor, that is, the entire figure. An example will clarify this. In the program `MeaslesNumber` start with $p = 0.1$ and $k = 2.5$. After about 30 iterations the program stops. From the 20th iteration on we see these numbers over and over again: ... 1.2250, 0.5359, 1.1577, 0.7012, 1.2250, 0.5359, 1.1577, 0.7012, ... etc. It is certainly not easy to understand why this happens, but from the definition it is undeniable that these four successive values determine the attractor for $k = 2.5$. The attractor is thus the set of those function values which emerge after a sufficiently large number of iterations. A set like that illustrated in Figure 2.2-2 is called a *strange attractor*.

In the region $k > 3$ there is just the attractor $-\infty$.

Whenever a function has several attractors, new questions are raised:

- Which regions of the k,p-plane belong to which attractor? That is, with which value p must I start, so that I am certain to reach a given objective - such as landing on the attractor 1?

- With which values should I start, if I do not wish to end at $-\infty$?

Because each sequence of numbers is uniquely determined, this question has a unique answer. Thus the k,p-plane is divided into clearly distinguished regions, whose boundaries are of considerable interest.

For the Feigenbaum diagram this problem can be solved in a relatively simple and clear fashion. But other cases, which we will encounter later, lead to surprising results.

For the above function

$$f(p) = p + k * p * (1 - p)$$

we can calculate the boundaries mathematically. Experimenting with the program `MeaslesNumber` should make it apparent that it is best to take negative p-values. Only then is there a chance of landing on the attractor.

This means that $f(p)$ must be > 0. From the equation (see Exercise 2.2.2-1 at the end of this section) this condition holds when

$$p < (k+1)/k.$$

Thus near $p = 0$ we have found the two boundaries for the catchment area, or *basin of attraction*, of the strange attractor. We will see that these boundaries do not always take such a smooth and simple form. And they cannot always, as in the Feigenbaum diagram, be described by simple equations. The problem of the boundaries between attracting regions, and how to draw these boundaries, will concern us in the next chapter.

In Figure 2.2.2-1 we again show the Feigenbaum diagram for the first quadrant of the coordinate system. We have superimposed the basin of attraction of the attractor.

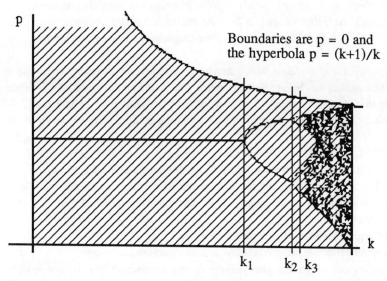

Boundaries are p = 0 and
the hyperbola p = (k+1)/k

Figure 2.2.2-1 Basin of attraction for the Feigenbaum diagram.

If you are interested in how the attractor looks and what its boundaries are when *k* is less than 0, try Exercise 2.2.2-2 at the end of this section.

Computer Graphics Experiments and Exercises for §2.2.2

Exercise   2.2.2-1
Show that $p+k*p*(1-p) > 0$, $p \neq 0$, $k \neq 0$ implies that $p < (k+1)/k$.

Exercise   2.2.2-2
So far we have described all phenomena in the case $k > 0$. What happens for $k \leq 0$ the reader/experimentalist must determine. To that end, three types of problem must be analysed:

• In which k,p-regions do we find stable solutions (that is, solutions not tending to $-\infty$)?

- What form does the attractor have?
- Where are the boundaries of the basins of attraction?

Exercise 2.2.2-3

We obtain a further extension of the regions to be examined, and hence extra questions to be answered, if we work with a different equation from that of Verhulst. One possibility is that we simplify the previous formula for $f(p)$ to

$$f(p) = k*p*(p-1).$$

This is just the term that describes the change from one generation to the next in the Verhulst equation. Investigate, for this example in particular, the basins of attraction and the value of δ.

Exercise 2.2.2-4

With enough courage, you can now try other equations. The results cannot be anticipated in advance, but they tend to be startling. Examples which provide attractive pictures and interesting insights are:

- $f(p) = k*p*p*(1-p)$ in the region $4 \le k \le 7$,
- $f(p) = k*p*(1-p*p)$ and other powers,
- $f(p) = k*\sin(p)*\cos(p)$ or square (cube, nth) root functions,
- $f(p) = k*(1 - 2*|p-0.5|)$ where $|x|$ means the absolute value of x.

2.2.3   Feigenbaum Landscapes

Even the simple fig-tree poses several puzzles. In the 'chaotic regime' we can see zones where points lie more thickly than in others. You can get a nice overview by representing the frequency with which the p-values fall inside a given interval. By putting the results together for different k-values, you will get a *Feigenbaum landscape* (Figures 2.2.3-1 and 2.2.3-2).

These Feigenbaum landscapes can be made to reveal further interesting structure. We suggest you experiment for yourself. It is naturally best if you develop the program yourself too, or at least try it out with your own parameter values. To help you in this task, we now provide some tips for the development of a Feigenbaum landscape program.

The appropriate range of values from 0 to 1.4 for $f(p)$ must be divided into a certain number of intervals. This number of course depends on the size of the screen display, which we have set using the constants Xscreen and Yscreen. In the program, for example, we have 280 'boxes', one for each interval.

For a given k-value the Feigenbaum program begins as usual. When a value of $f(p)$ falls within a given interval, this is noted in the corresponding box. After a sufficiently large number of iterations we stop the computation. Finally the results are displayed.

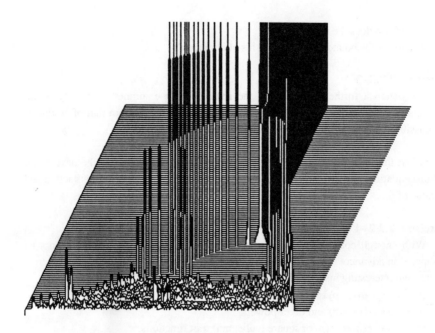

Figure 2.2.3-1   Feigenbaum landscape with the data 0, 3, 0, 1.4, 50, 500.

The box-number is counted from the right and the contents are drawn upwards, joining neighbouring values by a line.   The result is a curve resembling a mountain range, which describes the distribution of p-values for a given k-value.

To draw a picture with several k-values, we combine several such curves in one picture. Each successive curve is displaced two pixels upwards and one pixel to the right. In this way we obtain a 'pseudo-three-dimensional' effect. Of course the horizontal displacement can be to the left instead.   In Figure 2.2.3-3 ten such curves are drawn.

To improve the visibility, the individual curves must be combined into a unified picture. To achieve this, we do not draw the curves straight away.   Instead, for each horizontal screen coordinate (x-axis) we record in another field (in Pascal it is an *array*, just like the 'boxes') whichever of the previous vertical y-coordinates has the largest value. Only this maximal value is actually drawn.

With these hints you should be in a position to develop the program yourself.

A solution is of course given in §11.3.

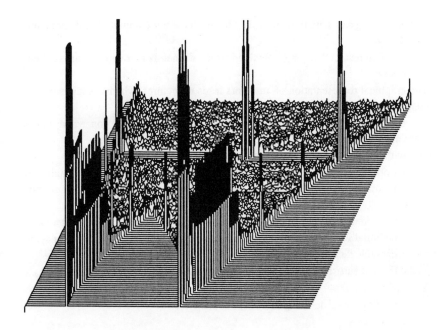

Figure 2.2.3-2   Feigenbaum landscape with the data 3, 2.4, 1.4, 0, 50, 500.

Computer Graphics Experiments and Exercises for §2.2.3

Exercise   2.2.3-1

Develop a program to draw Feigenbaum landscapes.   Use the resulting 'three-dimensional measuring instrument' to investigate interesting sections of the Feigenbaum diagram.   We have already given hints for the main steps above.

Exercise   2.2.3-2

Generalise the pseudo-three-dimensional landscape method, so that other formulas can be represented in the same way.

2.3   Chaos – Two Sides to the Same Coin

In the previous chapter you were confronted with many new concepts. Furthermore, your own experiments will certainly have given you more to think about, so that the basic idea of the first chapter may have been somewhat obscured.   We will therefore discuss the initial consequences of our investigations, before we embark on new adventures. What have we discovered?

- The interesting cases are those in which the results of our computations do not tend to ∞ or −∞.
- The set of all results that can be obtained after sufficiently many iterations is called an *attractor*.
- The graphical representation of attractors leads to pictures, which contain smaller copies of themselves.
- Three new concepts - *attractor*, *frontier*, and *bifurcation* - are connected with these mathematical features.

We began with the central idea of 'experimental mathematics':

- These important concepts in the theory of dynamical systems are based on taking an arbitrary mathematical equation and 'feeding it back' its own results again and again.

By choosing different starting values we repeatedly find the same results upon iteration. With the same initial values we always obtain the same results. There are however some deep and remarkable facts to be observed:

In the Feigenbaum diagram we can distinguish three regions:

- $k \leq 2$ (*Order*);
- $2 < k < 2.57$ (*Period doubling cascade*: $0 \leq p \leq 1.5$);
- $k \geq 2.57$ (*Chaos*).

Under certain conditions, moreover, we cannot predict what will happen at all. Insignificant differences in the initial value lead to totally different behaviour, giving virtually unpredictable results. This 'breakdown of computability' happens around $k = 2.57$. This is the 'point of no return', dividing the region of order from that of chaos.

To avoid misunderstandings, we must again emphasise that the above remarks refer to a completely deterministic system. But - from a practical point of view - the chaos effect produces the bitter aftertaste of indeterminacy.

Mathematicians try to find models that can describe the 'long-term behaviour' of a system. The Feigenbaum scenario exemplifies the behaviour of the simplest nonlinear system. The message is that any nonlinear system may exhibit similar phenomena. Complex systems, depending on many parameters, can under certain conditions switch from stable conditions to instability. We speak of chaos.

Of course we want to keep on the track of this essentially philosophical question. It seems that there is some deep connection between order and chaos, which we cannot yet make explicit.

One thing is certain.

As a result of our previous investigations, order and chaos are two sides to the same coin - a parameter-sensitive classification.

Enough theory!

In the following chapter we will return to the question. But now you must try out some exercises for yourself. Good luck with your experiments!

3 Strange Attractors

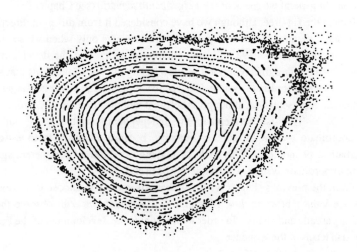

3.1   The Strange Attractor

Because of its aesthetic qualities, the Feigenbaum diagram has acquired the nature
of a symbol.   Out of allegedly dry mathematics, a fundamental form arises.   It describes
the connection between two concepts, which have hitherto seemed quite distinct: order
and chaos, differing from each other only by the values of a parameter.   Indeed the two
are opposite sides of the same coin.   All nonlinear systems can display this typical
transition.   In general we speak of the *Feigenbaum scenario* (see Chapter 9).

Indeed the fig–tree, although we have considered it from different directions, is an
entirely static picture.   The development in time appears only when we see the picture
build up on the screen.   We will now attempt to understand the development of the
attractor from a rather different point of view, using the two dimensions that we can set
up in a cartesian coordinate system. The feedback parameter k will no longer appear in
the graphical representation, although as before it will run continuously through the range
$0 \leq k \leq 3$.   That is, we replace the independent variable k in our previous
(k,p)-coordinate system by another quantity, because we want to investigate other
mathematical phenomena.   This trick, of playing off different parameters against each
other in a coordinate system, will frequently be useful.

From the previous chapter we know that it is enough to choose k between 0 and 3.
There are values between $k = 1.8$ and $k = 3$ at which we can observe the period-
doubling cascade and chaos.   In order to investigate the development of the Feigenbaum
diagram in terms of the sequence

$$p_{n+1} = p_n + k * p_n * (1 - p_n),$$

we choose as coordinate system the population values p_n and p_{n+1} which follow each
other in the sequence.   To the right we draw the final value p_n of the previous iteration,
and we draw the result $p_{n+1} = f(p_n)$ vertically.   We know this construction already
from graphical iteration (see Chapter 2.1.2).

If you have already set up the program `Feigenbaum` then the modifications
required are relatively easy. They relate solely to the part that does the drawing. Instead
of the coordinates (k,p) we must now display $(p, f(p,k))$ on the screen.   In the
program fragment only the following part changes:

```
FOR i = 0 to Visible DO
BEGIN
    SetUniversalPoint (population, f (population, Feedback));
    population := f (population, Feedback);
END;
```

Nothing else need be altered.

You should make this modification to your existing program and see what happens.
The final result (Figure 3.1-1) can only convey an incomplete impression of the
dynamical development that occurs during its generation. You are advised to observe the
gradual growth of this figure on the screen. If we choose the same scale on both axes the
picture begins (for k a little larger than 0) in a less than spectacular way. The diagonal

straight line, which first appears, expresses the fact that the underlying value is tending to a constant. Then $p = f(p)$. After $k = 2$ we obtain two alternating underlying values.

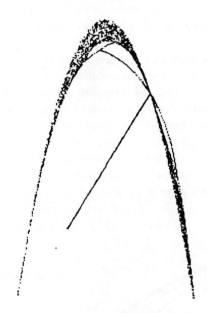

Data: 0, 3, 0, 1.4, 50, 50 for $0 \leq k \leq 3$

Figure 3.1-1 'Trace' of the parabola-attractor in the $p, f(p)$-plane.

The figure grows in two directions. Low starting values for the formula produce a higher result and then return. The curious picture here has the form of a thin curved line and runs roughly perpendicular to the original bisector. For periods 4 and 8 – when the figure grows in 4 or 8 places – it is also easy to see how the starting value p and the result $f(p)$ are connected. Thus we have built ourselves yet another measuring or observing instrument, with which we can watch the temporal development of period-doubling. As soon as we enter the chaotic region, a well-known mathematical object appears: the parabola.

If you want to draw this and similar pictures, please take a look at Exercises 3-1 and 3-2 at the end of this chapter. However, you will need a certain amount of patience, because in Figure 3.1-1 it takes some time, after the diagonal line is drawn, before points scatter on to the parabola.

In order to delve more deeply into the 'history' of the sequence, it is necessary to link together the results not just of one, but of several previous values. The investigation of the so-called Verhulst attractor (Figure 3.1-2) is especially interesting. This is the attractor corresponding to the equation

```
f(pn)  = pn + 1/2*k*(3*pn*(1-pn)-pnMinus1*(1-pnMinus1))
```
which we have already encountered in §2.2.

For small k-values we soon run into a boundary, which of course lies on the diagonal of the $(p,f(p))$-coordinate system. We can observe the periods 2, 4, 8, etc. When we reach the chaos-value $k=1.6$ things get very exciting.   At the first instant a parabola suddenly appears. Furthermore, it is not evenly filled, as we have seen already. It has an 'internal structure'.

Let us once more collect our conclusions together.   The geometric form of the attractor arises because we draw the elements of a sequence in a coordinate system.   To do this we represent the starting value p in one direction, and in the other the result $f(p)$ of the iteration for a fixed value of k.   We first notice that the values for $f(p)$ do not leave a certain range between 0 and 1.4.   Furthermore, we notice genuine chaos, revealing either no periodicity at all or a very long period.   Here we know that more and more points appear in a completely unpredictable way.   These points form lines or hint at their presence. Under careful observation the attractor seems to sit on a parabolic curve, defined by numerous thin lines. We want to take a closer look at that!

Figure 3.1-2   The Verhulst attractor for $k = 1.60$.

The changes in our previous program that are needed to generate the Verhulst attractor are again very simple:

Program Fragment 3.1-1

```
(* BEGIN: problem-specific procedures *)
PROCEDURE VerhulstAttractor;
    VAR i : integer;
```

```
            pn, pnMinus1, pnMinus2, oldValue: real;

FUNCTION f(pn : real) : real;
BEGIN
    pnMinus1 := pnMinus2;
    pnMinus2 := pn;
    f := pn + Feedback/2*(3*pn*(1-pn) -
           pnMinus1*(1-pnMinus1));
END;

BEGIN
    pn := 0.3; pnMinus1 := 0.3; pnMinus2 := 0.3;
    FOR i := 0 TO Invisible DO
        pn  := f(pn);
    REPEAT
        oldValue := pn;
        pn := f(pn);
        SetUniversalPoint (pn, oldValue);
    UNTIL Button;
END;
```

The value of Feedback is constant during each run of the program, e.g. $k = 1.6$. In order to experiment with different k-values, the variable Feedback - and also Left, Right, Bottom, Top - must be input.

Using this VerhulstAttractor program we draw, in the first instance, the whole attractor, when p and $f(p)$ lie between 0 and 1.4 (see Exercise 3-3). If we choose a different range of values, we get different sections of it. To begin with, we look at places where there seems to be just a line, and then at the 'nodes' where the 'lines' meet or cross. These lines break up when they are magnified. In fact, they are really 'chains' into which the points arrange themselves. The picture resembles an aerial photograph of a large number of people walking in the snow along pre-defined tracks. The starting-point and destination are nowhere to be seen. By looking closely enough we can distinguish a faint track (along which the points/people lie more thickly) and parallel to it a wider one, on which the points are distributed more irregularly. A magnification of the wider track shows exactly the same structure. The same even happens if we magnify the thin track by a larger amount. If we examine the 'nodes' more carefully, we obtain something we have already encountered: a smaller version of the same attractor. This remarkable behaviour has already arisen in the Feigenbaum diagram. Many sections of the diagram produce the entire figure. This is the phenomenon of self-similarity again.

The strange Verhulst attractor exhibits a structure assembled from intricate curves

(Figure 3.1-3 ff.).   By looking closely enough we encounter the same structure
whenever we magnify the attractor. This structure is repeated infinitely often inside itself
and occurs more often the more closely we look. The description of this 'self-similarity',
and the aesthetic structure of the Verhulst attractor already referred to, are developed
further in exercises at the end of the chapter.

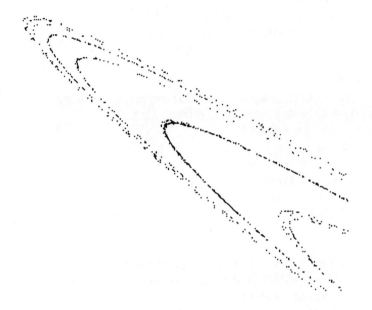

Figure 3.1-3   Self-similarity at each step (top left projections of the attractor).

As already mentioned at the end of §2.1, we have wandered some way from our
original problem ('measles in the children's home'). We will now explain the
mathematical background and some possible generalisations. The equation on which the
Verhulst attractor is based is well known as a numerical procedure for the solution of a
differential equation. By this we mean an equation in which a function y and one of
more of its derivatives occur, either linearly or nonlinearly. You remember: the first
derivative y' describes how rapidly y changes. The second derivative y'' describes
how y' changes, and hence the curvature of y. The simplest form for a nonlinear
differential equation is

$$y' = y*(1-y) = g(y).$$

The symbol $g(y)$, which we use for $y*(1-y)$, will simplify the later description. A
nonlinear differential equation is an equation in which the function y occurs
quadratically, to a higher power, or for example as a trignometric expression.

Numerical methods are known for solving such equations. Starting from an initial value y_0, we try to approximate the equation in small steps. The rapidity with which we approach the limiting value (if one exists) is represented by the symbol k.

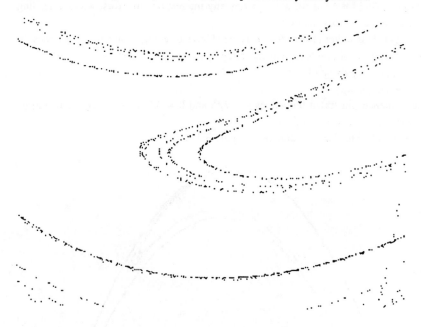

Figure 3.1–4 Self–similarity at each step (detail from the 'node' to the right of centre).

The simplest technique, known as *Euler's method*, is described e.g. in Abramowitz and Stegun (1968):

$$y_{n+1} = y_n + k*y'_n + O(k^2).$$

The final term expresses the fact that the equation is not exact, and that the error is of the order of magnitude of k^2. Since we have previously considered many iterations, the error interests us no further. The estimate is simplified if instead of $y'_n = g(y_n) = y_n*(1-y_n)$ we substitute

$$y_{n+1} = y_n + k*y'_n = y_n + k*g(y_n) = y_n + k*y_n*(1-y_n).$$

Thus we have recovered our old friend, the Feigenbaum formula, from § 2.2!

There now opens up a promising approach to interesting graphical experiments: choose a differential equation that is easy to compute, and try to approximate the solution by a numerical method. The numerical method, in the form of an iterative procedure, can then be taken as the basis of the graphical experiment.

In the same way we can derive the equation for the Verhulst attractor. The starting-point is the so-called *two-step Adams-Bashforth method*. It is somewhat more complicated than the Euler method (see Exercises 3-6 and 3-7).

3.2   The Hénon Attractor

We can find formulas for other graphically interesting attractors, without needing any special mathematical background.

Douglas Hofstadter (1981) describes the Hénon attractor.   On page 7 he writes: 'It refers to a sequence of points (x_n, y_n) generated by the recursive formulas

$$x_{n+1} = y_n - a*x_n^2 + 1$$
$$y_{n+1} = b*x_n.$$

For the sequence illustrated the values $a = 7/5$ and $b = 3/10$ were taken; the starting values were $x_0 = 0$ and $y_0 = 0$.'

Contemplate the Hénon attractor of Figure 3.2-1.

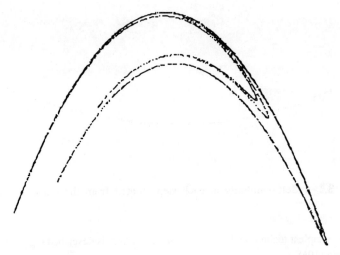

Figure 3.2-1 The Hénon attractor

Like the Feigenbaum diagram, the Hénon attractor should not be thought of as just a mathematical toy which produces remarkable computer graphics.   In 1968, Michel Hénon, at the Institute for Astrophysics, Paris, proposed taking such simple quadratic mappings as models, to carry out computer-graphical simulations of dynamical systems. In particular, he was thinking of the study of the orbits of asteroids, satellites, and other heavenly bodies, or of electrically charged particles in particle accelerators.

During the period 1954–63 the mathematicians Kolmogorov, Arnold, and Moser developed a theory centred around the so-called KAM theorem.   In it, they studied the behaviour of a stable dynamical system – such as, for example, a satellite circling the Earth – to clarify what happens when tiny external forces act on it. Planets or asteroids which orbit round the Sun often undergo such perturbations, so that their orbits are not truly elliptical. The KAM theorem attempts to decide whether small perturbations by external forces can lead to instability – to chaos – in the long-term behaviour.   For

instance an asteroid can be disturbed in its path by the gravitational force of Jupiter.
Here one talks of *resonance*. Such resonances occur when the ratio of the orbital periods
is a rational number. If, for example, two of Jupiter's orbits take the same time as five
orbits of an asteroid, we have the case of a *2:5 resonance*.

In the title picture of Chapter 3 we see (upper figure) such a computer simulation,
where in the course of the simulation ever stronger external influences are imposed. The
inner curves elucidate the influence on the orbital behaviour of small external
disturbances. Each point in the drawing shows the position of the asteroid after a further
revolution. For small disturbances only small differences can be seen, and the system
remains stable. When the influence of the external disturbance increases, we observe six
'islands'. These represent a *1:6 resonance*. A body, such as for instance an asteroid,
having 1/6 the orbital period of Jupiter, would find itself in such a *resonance band*.

Further out, we see dotted regions of instability. The behaviour of an asteroid in
this region is no longer predictable, because small external influences can have large
effects. It is even possible for the asteroid to be catapulted out of its orbit into the
'emptiness' of the universe. Scientists believe that the gaps in the asteroid belt can be
explained by this mechanism.

This brief explanation should make clear the connection between such simple
formulas, and deep effects in the field of macroscopic physics. Further explanation can
be found in physics texts or in Hughes (1986).

The formula to generate the title picture of Chapter 3 is as follows:
$$x_{n+1} = x_n * \cos(w) - (y_n - x_n^2) * \sin(w)$$
$$y_{n+1} = x_n * \sin(w) - (y_n - x_n^2) * \cos(w).$$
Here w is an angle in the range $0 \leq w \leq \pi$.

Compare the structure of this formula with the one described by Hofstadter at the
start of this chapter. A program fragment for generating (other) Hénon attractors is as
follows:

Program Fragment 3.2–1

```
PROCEDURE HenonAttractors;
(* x0, y0, dx0, dy0 global variables *)
    VAR
        cosA, sinA : real;
        xNew, yNew, xOld, yOld : real;
        deltaxPerPixel, deltayPerPixel : real;
        ok1, ok2 : boolean;
        i, j : integer;
BEGIN
    cosA := cos (phaseAngle); sinA := sin (phaseAngle);
    xOld := x0; yOld := y0;  {starting point of first orbit}
    deltaxPerPixel := Xscreen/(Right-Left);
    deltayPerPixel := Yscreen/(Top-Bottom);
```

```
FOR j = 1 to orbitNumber DO
BEGIN
   i := 1;
   WHILE i <= pointNumber DO
   BEGIN
      IF (abs(xOld)<= maxReal) AND (abs(yOld)<= maxReal)
      THEN
      BEGIN
         xNew := xOld*cosA - (yOld - xOld*xOld)*sinA;
         yNew := xOld*sinA + (yOld - xOld*xOld)*cosA;
         ok1 := (abs(xNew-Left) < maxInt/deltaxPerPixel);
         ok2 := (abs(Top-yNew) < MaxInt/deltayPerPixel);
         IF ok1 AND ok2 THEN
         BEGIN
            SetUniversalPoint (xNew, yNew);
         END;
         xOld := xNew;
         yOld := yNew;
      END;
      i := i+1;
   END;    {WHILE i}
   xOld := x0 + j * dx0;
   yOld := y0 + j * dy0;
END;    {FOR j := ...}
END;
(* END : problem-specific procedures *)
```

3.3   The Lorenz Attractor

Five years before Michel Hénon began working in Paris on models for simulating dynamical systems, equally exciting things were happening elsewhere. In 1963, working in a completely different area, the American Edward N. Lorenz wrote a remarkable scientificarticle.

In his article Lorenz described a family of three particular differential equations[1] with parameters a, b, c :

$$x' = a*(y-x)$$
$$y' = b*x-y-x*z$$
$$z' = x*y-c*z.$$

Numerical analysis on a computer revealed that these equations have extremely

---

[1]We write the first derivatives with respect to time as x', y', z', etc. For example x' will be written in place of dx/dt.

complicated solutions. The complicated connections with and dependences upon the parameters could at first be elucidated only through computer-graphical methods.

The interpretation of the equations was exciting. In particular Lorenz sought - and found - a mathematical description which led to a rational explanation of the phenomenon of unpredictability of the weather in meteorology.

The idea of the model is as follows. The Earth is heated by the Sun. Part of the energy received at the Earth's surface is absorbed and heats the atmosphere from below. From above, the atmosphere is cooled by radiation into space. The lower, warmer layers of air want to rise upwards, and the upper, colder layers want to fall downwards. This transport problem, with oscillating layers of cold and warm air, can lead to turbulent behaviour in the atmosphere.

The picture's remarkable appearance cannot fully capture the surprising behaviour that occurs while it is being drawn on the screen. In consequence it is very important that you program and experiment for yourself.

Having said that, let us look first at some pictures of the Lorenz attractor.

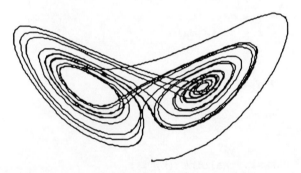

Figure 3.3-1 Lorenz attractor for $a = 10$, $b = 28$, $c = 8/3$, and screen dimensions -30, 30, -30, 80.

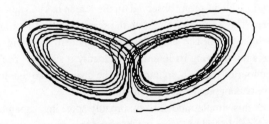

Figure 3.3-2 Lorenz attractor for $a = 20$, $b = 20$, $c = 8/3$, and screen dimensions -30, 30, -30, 80.

The program fragment that generates the figure is like this:

Program Fragment 3.3-1

```
(* START : Problem-specific procedures *)
PROCEDURE LorenzAttractor;
    VAR x, y, z : real;
    PROCEDURE f;
      CONST
          delta = 0.01;
      VAR
          dx, dy, dz : real;
    BEGIN
      dx := 10*(y-x);
      dy := x*(28-z)-y;
      dz := x*y - (8/3)*z;
      x := x + delta*dx;
      y := y + delta*dy;
      z := z + delta*dz;
    END;
BEGIN
    x := 1; y := 1; z := 1;
    f;
    SetUniversalPoint (x,z);
    REPEAT
      f;
      DrawUniversalLine (x, z);
    UNTIL Button;
END;
(* END : Problem-specific procedures *)
```

The behaviour which Lorenz observed on the screen in 1963 can be described in the following manner. The wandering point on the screen circles first round one of the two foci around which the two-lobed shape develops. Suddenly it changes to the other side. The point wanders on, drawing its line, until suddenly it switches back to the other side again. The behaviour of the path, in particular the change from one lobe to the other, is something that we cannot predict in the long run.

Even though this simple model is not, broadly speaking, capable of explaining the complex thermodynamic and radiative mechanisms that go on in the atmosphere, it does establish two points:

• It illustrates the basic impossibility of precise weather-prediction. Lorenz talked about this himself, saying that the 'fluttering of a butterfly's wing' can influence the weather. He called it the *butterfly effect*.

• It raises the hope that very complex behaviour might perhaps be understood through simple mathematical models.

This is a basic assumption, from which scientists in the modern theory of dynamical systems start. If this hypothesis had been wrong, and had not on occasion already proved to be correct, everybody's scientific work in this area would have been pointless.

As far as we are concerned, it remains true that certain principles on the limits to predictability exist, which cannot be overcome even with the best computer assistance.

To understand and pin down this phenomenon is the aim of scientists in the borderland between experimental mathematics, computer graphics, and other sciences.

Computer Graphics Experiments and Exercises for Chapter 3

Exercise 3-1

Modify the program `Feigenbaum` so that you can use it to draw the parabola attractor. Investigate this figure in fine detail. Look at important k-intervals, for example those with a uniform period. Look at the region near the vertex of the parabola, magnified.

Exercise 3-2

Carry out similar investigations using the number sequences described above in §2.2.

Exercise 3-3

Starting from `Feigenbaum` develop a Pascal program to draw the Verhulst attractor with the value $k = 1.6$, within the ranges $0 \leq p \leq 1.4$ and $0 \leq f(p) \leq 1.4$. Start with $p = 0.3$ and do not draw the first 20 points. Compare your result with Figure 3.1-2.

Investigate different sections from this figure, in which you define the boundaries of the drawing more closely.

Exercise 3-4

Make an animated 'movie' in which several pictures are shown one after the other. The pictures should show a section of the attractor at increasingly large magnification. We recommend a region near the 'node' with coordinates $p = 0.6$, $f(p) = 1.289$. Start with the entire picture. In this connection we offer a warning: the more extreme the magnification, the more points you must compute that lie outside the region being drawn. It can take more than an hour to put the outline of the attractor on the screen.

Exercise   3-5

Make a 'movie' in which a sequence of magnified sections of the attractor are superimposed on each other.

Exercise   3-6

The two-step Adams–Bashforth method can be written like this:
$$y_{n+1} = y_n + \tfrac{1}{2}*k*(3*g(y_n)-g(y_{n-1})).$$
If we substitute $g(y) = y*(1-y)$ we obtain
$$f(y_n) = y_n+\tfrac{1}{2}*k*(3*y_n*(1-y_n)-y_{n-1}*(1-y_{n-1})).$$
If we use the current variable p and the previous value pnMinus1, we get the familiar formula

```
f(pn) = pn+1/2*k*(3*pn*(1-pn)-pnMinus1*(1-pNminus1)).
```

Try out the above method to find solutions of differential equations, and also other variations on the method, such as:
$$y_{n+1} = y_{n-1} + 2*k*g(y_n)$$
$$y_{n+1} = y_n+ k/2*(g(y_n)+g(y_{n-1}))$$
$$y_{n+1} = y_n+k/24*(55*g(y_n)-59*g(y_{n-1})+37*g(y_{n-2})-9*g(y_{n-3})$$
or whatever else you can find in your mathematical textbooks.

Calculate Feigenbaum diagrams and draw the attractor in $(p,f(p))$-coordinates.

Exercise 3.7

We will relax the methods for solving differential equations further. The constants 3 and –1 appearing in the Adams–Bashforth method are not sacred. We simply change them.

Investigate the attractors produced by the recursion formula

```
f(p) = p + 1/2*k*(a*p*(1-p)+b*pnMinus1*(1-pnMinus1)).
```

Next work out, without drawing anything, which combinations of a and b can occur without the value $f(p)$ becoming too large. Put together a 'movie' of the changes in the attractor, occurring when the parameter a alone is varied from 2 to 3, with $a = 3, b = -1$ as the end point.

Exercise   3-8

Write a program to draw the Hénon attractor. Note that you do not have to use the same scale on each coordinate axis. Construct sections of this figure. Change the values of a and b in the previous exercises. Get hold of the article by Hofstadter (1981) and check his statements.

Exercise   3-9

Experiment with the system of equations for the 'planetary' Hénon attractor. Data for the title picture, for instance, are:

```
phaseAngle := 1.111; Left := -1.2, Right := 1.2;
```

```
Bottom := -1.2; Top := 1.2; x0 := 0.098; y0 := 0.061;
dx0 := 0.04; dy0 := 0.03; orbitNumber := 40;
pointNumber := 700;
```
further suitable data can be found in Hughes (1986).

Exercise 3-10

Experiment with the Lorenz attractor. Vary the parameters a, b, c. What influence do they have on the form of the attractor?

Exercise 3-11

Another attractor, called the *Rössler attractor*, can be obtained from the following formulas:

$$x' = -(y+z)$$
$$y' = x+(y/5)$$
$$z' = 1/5 + z*(x-5.7).$$

Experiment with this creature.

4 Greetings from Sir Isaac

In the previous chapters we saw what the 140–year–old Verhulst formula is capable of when we approach it with modern computers. Now we pursue the central ideas of self–similarity and chaos in connection with two further mathematical classics. These are *Newton's method* for calculating zeros, and the *Gaussian plane* for representing complex numbers.

In both cases we are dealing with long–established methods of applied mathematics. In school mathematics both are given perfunctory attention from time to time, but perhaps these considerations will stimulate something to change that.

4.1   Newton's Method

A simple mathematical example will demonstrate that chaos can be just around the next corner.

Our starting point is an equation of the third degree, the cubic polynomial

$$y = f(x) = (x+1)*x*(x-1) = x^3-x.$$

This polynomial has zeros at $x_1 = -1$, $x_2 = 0$, and $x_3 = 1$ (Figure 4.1-1).

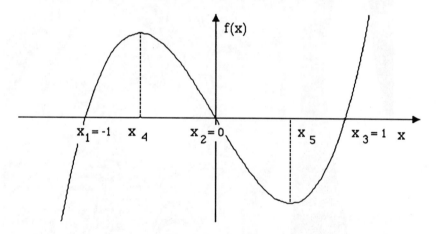

Figure 4.1-1 The graph of the function $f(x) = (x+1)*x*(x-1)$.

In order to introduce chaos into the safe world of this simple mathematical equation, we will apply the apparently harmless Newton method to this function.

Sir Isaac Newton was thinking about a widely encountered problem in mathematics: to find the zeros of a function, for which only the formula is known. For equations of the first and second degree we learn simple methods of solution at school, and complicated and tedious methods are known for polynomials of degree 3 or 4. For degree 5 there is no simple expression for the solution in closed form.   However, complicated equations like these, and others containing trigonometric or other functions, are of interest in many applications.

Newton's point of departure was simple: find the zeros by trial and error. We start with an arbitrary value, which we will call x_n. From this the function value $f(x_n)$ is calculated. In general we will not have found a zero, that is, $f(x_n) = 0$. But from here we can 'take aim' at the zero, by constructing the tangent to the curve. This can be seen in Figure 4.1-2. When constructing the tangent we need to know the slope of the curve. This quantity is given by the derivative $f'(x_n)$, which can often be found easily.[1]

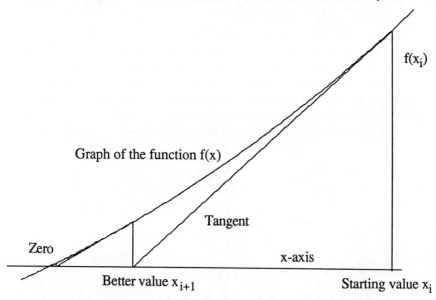

$f(x_i)$

Graph of the function f(x)

Tangent

Zero

x-axis

Better value x_{i+1}

Starting value x_i

Figure 4.1-2 How Newton's method leads to a zero

A further problem should be mentioned. If $f'(x_n) = 0$, we find ourselves at a minimum, maximum, or inflexion point. Then we must carry out the analysis at the point $x_n + dx$.

The right-angled triangle in Figure 4.1-2 represents a slope of height $f(x_n)$ and width $f(x_n)/f'(x_n)$. From this last expression we can correct our approximate zero x_n, to get a better value for the zero:

$$x_{n+1} = x_n - f(x_n)/f'(x_n).$$

An even better approximation arises if we carry out this calculation again with x_{n+1} as input.

Basically Newton's method is just a feedback scheme for computing zeros.

---

[1]Even when $f'(x_n)$ is not known as an explicit function we can approximate the differential quotient closely, by $f'(x_n) = (f(x_n+dx)-f(x_n-dx))/(2*dx)$, where dx is a small number, e.g. 10^{-6}.

When we get close enough to the zero for our purposes, we stop the calculation. A criterion for this might be, for example, that f(x) is close enough to zero ($|f(x_n)| \leq 10^{-6}$) or that x does not change very much ($|x_n - x_{n+1}| \leq 10^{-6}$).

For further investigation we return to the above cubic equation

$$f(x) = x^3 - x$$

for which

$$f'(x) = 3x^2 - 1.$$

Then we compute the improved value x_{n+1} from the initial value x_n using

$$x_{n+1} = x_n - (x_n{}^3 - x_n)/(3x_n{}^2 - 1).$$

If we exclude[2] the two irrational critical points $x_4 = -\sqrt{(1/3)}$ and $x_5 = \sqrt{(1/3)}$, then from any starting value we approach one of the three zeros x_1, x_2, or x_3. These zeros thus have the nature of attractors, because every possible sequence of iterations tends towards one of the zeros. This observation leads to a further interesting question: given the starting value, which zero do we approach?   More generally: what are the basins of attraction of the three attractors x_1, x_2, and x_3?

The first results of this method will be shown in three simple sketches.

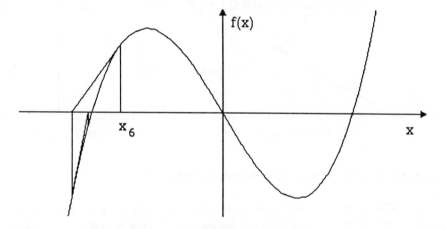

Figure 4.1-3 Initial value x_6 leads to attractor x_1.

We know the position of the axes and the graph of the function. At each iteration, we draw a vertical line and construct the tangent to the curve at that point. The result of Figures 4.1-3 to 4.1-5 is not particularly surprising: if we begin with values x_6, x_7, or x_8 close to an attractor, the the iteration converges towards that same attractor.

By further investigation we can establish:

---

[2]These numbers cannot be represented exactly in the computer. The Newton method fails here because the first derivative $f'(x_n) = 0$. Graphically, this follows because the tangents at these points are horizontal, and obviously cannot cut the x-axis, because they run parallel to it.

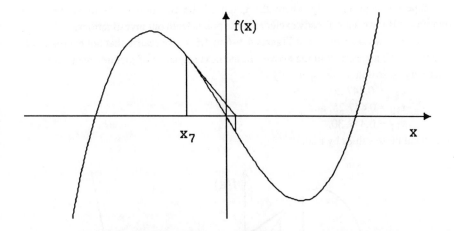

Figure 4.1-4   Initial value x_7 leads to attractor x_2.

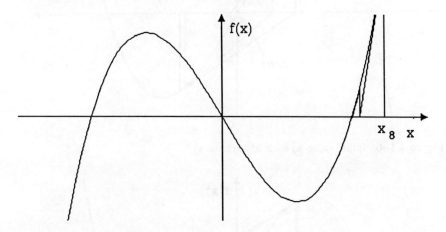

Figure 4.1-5   Initial value x_8 leads to attractor x_3.

- The basin of attraction of the attractor x_1 includes the region
  $-\infty < x < x_4 = -\sqrt{(1/3)}$.
- The basin of attraction of the attractor x_3 includes the region
  $x_5 = \sqrt{(1/3)} < x < \infty$.
  In particular, this region is symmetrically placed relative to the basin of attraction of x_1.
- The numbers near the origin belong to the basin of attraction of x_2.
- If we have found the attractor for a given initial value, then nearby initial values lead to the same attractor.

We expect exceptions only where the graph of the function has a maximum or a minimum. But we have already excluded these points from our investigations.

If we now take a glance at Figures 4.1-6 to 4.1-8, we realise that not everything is at simple as it appears from the above. In the next sequence of pictures we begin from three very close initial values, namely

$x_9 \quad = 0.447\ 20,$

$x_{10} = 0.447\ 25,$ and

$x_{11} = 0.447\ 30.$

We found these values by trial.

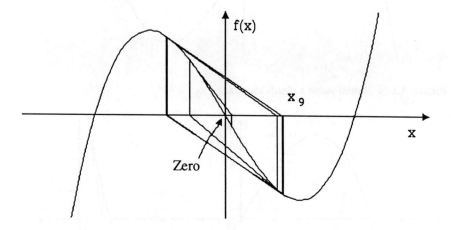

Figure 4.1-6 Initial value x_9 leads to attractor x_2.

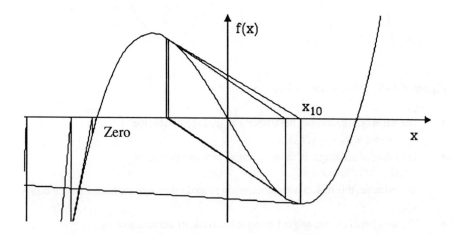

Figure 4.1-7 Initial value x_{10} leads to attractor x_1.

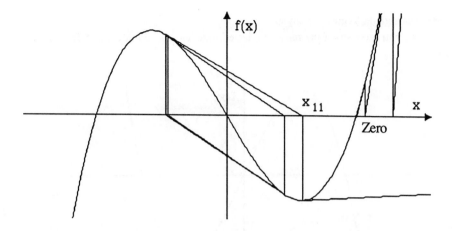

Figure 4.1-8 Initial value x_{11} leads to attractor x_3.

Despite their closeness, and despite the smooth and 'harmless' form of the graph of the function, the Newton method leads to the three different attractors. A sensible prediction seems not to be possible here. We refer to this 'breakdown of predictability' when we speak below of chaos.

In all areas of daily life, and also in physics and mathematics, we make use of a great number of unspoken assumptions, when we describe things or processes. One of the basic principles of physics is the causality principle.[3] Recall that this states that *the same causes* lead to *the same effects*.   If this rule did not hold, there would be no technical apparatus upon which one could rely.   Interestingly, this precept is often handled in a very cavalier fashion.   We formulate this generalisation as the strong causality principle: *similar causes* lead to *similar effects*.

That this statement does not hold in general is obvious every Saturday in Germany when the lottery numbers are called – the result is not similar, even though the 49 balls begin each time in the same (or at least a similar) arrangement. Our definition of chaos is no more than this:

A chaotic system is one in which the strong causality principle is broken.

In the next step – and indeed in the whole book – we will show that such chaos is not totally arbitrary, but that at least in some regions an order, at first hard to penetrate, lies behind it.

In order to explain this order, in Figure 4.1-9 we have drawn the basins of attraction for the three attractors of the function $f(x)$ in different shades of grey and as rectangles of different heights. Everywhere we find a short, medium–grey rectangle, the iteration tends towards the attractor x_1. The basin of attraction of x_2 can be recognised as the light–grey, medium–height rectangles; and all points which tend towards x_3 can

---

[3]Causality: logical consequence.

be seen as high dark-grey rectangles.

Can you reconstruct the results collected together after Figure 4.1-5 from Figure 4.1-9?

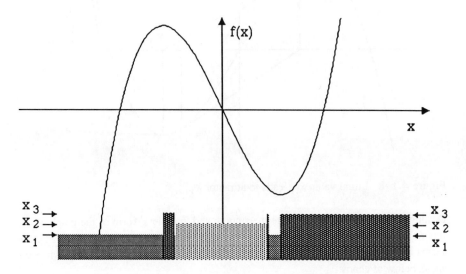

Figure 4.1-9 Graphical representation of the basins of attraction.

Especially interesting are the *chaotic zones*, in which there is a rapid interchange between the basins (of attraction). We show a magnified version of the left-hand region (for *x*-values in the range $-0.6 < x < -0.4$) in Figure 4.1-10. The section from Figure 4.9-9 is stretched along the *x*-axis by a factor of 40. The graph of the function in this range is barely distinguishable from a straight line.

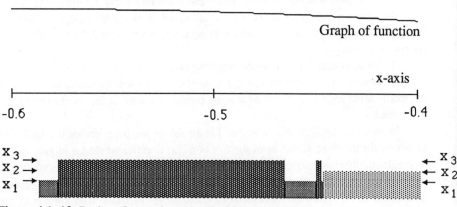

Figure 4.1-10 Basins of attraction (detail of Figure 4.1-9).

If we now investigate the basins shown by the grey areas we observe the following:
- On the outside we find the basins of x_1 and x_2.
- A large region from the basin of x_3 has sneaked between them.
- Between the regions for x_3 and x_2 there is another region for x_1.
- Between the regions for x_1 and x_2 there is another region for x_3.
- Between the regions for x_3 and x_2 there is another region for x_1,

and so on.

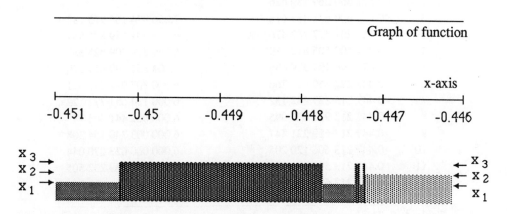

Figure 4.1–11 Basins of attraction (detail of Figure 4.1–10).

A further magnification by a factor of 40 in Figure 4.1–11 shows the same scheme again, but on a finer scale. We have already met the scientific description of this phenomenon: self-similarity.

The apparent chaos reveals itself as a strongly ordered zone.

In a further investigation we will now set to work, calculating as closely as possible the points that separate the basins from each other. The corresponding program will not be described further here: it is left as an exercise. These boundary points will be denoted g_i. Only the results are shown, in Table 4–1.

- The first value is given by $g_1 = -0.577\,35\dots$.
- If $x < g_1$ then x belongs to the basin of x_1.
- If $g_1 < x < g_2$ then x belongs to the basin of x_3.
- If $g_2 < x < g_3$ then x belongs to the basin of x_1, and so on.

Using Table 4–1 we have discovered a simple mathematical connection between the g_i-values. Namely, the quotient tends to a constant value. In fact

$$\lim_{n \to \infty} \frac{g_n - g_{n-1}}{g_{n+1} - g_n} = 6.0 = q \ .$$

Index n	g_n	$(g_n - g_{n-1})/(g_{n+1} - g_n)$
1	-0.577 350 269 189 626	–
2	-0.465 600 621 433 678	7.256 874 166 975 182
3	-0.450 201 477 782 476	6.179 501 149 801 554
4	-0.447 709 505 812 910	6.029 219 709 825 583
5	-0.477 296 189 979 436	6.004 851 109 839 370
6	-0.447 227 359 657 766	6.000 807 997 292 021
7	-0.447 215 889 482 132	6.000 134 651 772 122
8	-0.447 213 977 829 095	6.000 022 441 303 783
9	-0.447 213 659 221 447	6.000 003 740 154 308
10	-0.447 213 606 120 205	6.000 000 623 270 044
11	-0.447 213 597 269 999	6.000 000 039 232 505
12	-0.447 213 595 794 965	–

Table 4-1 The basin boundaries

For x-values greater than zero the same result holds but now with positive g_i-values. The resulting quotient is the same.

A few words to explain these numbers are perhaps in order.

- The number g_1 has the value $g_1 = \sqrt{(1/3)} = 3^{-1/2}$.

This is worked out by applying school methods to investigate the curve

$$f(x) = x^3 - x.$$

At $x = g_1$ the first derivative

$$f'(x) = 3x^2 - 1$$

takes the value $f'(g_1) = 0$.

The function $f(x)$ has an extreme value there, at which the derivative changes sign, so an increasing function becomes decreasing. Because the slope (first derivative) plays a special role in Newton's method, this leads us to conclude that the points on the right and left of an extreme value belong to distinct basins of attraction.

- For the limiting value of the g_i we have:

$$\lim_{n \to \infty} g_n = \sqrt{\frac{1}{5}} = x_g \ .$$

This value too can be expressed analytically.[4]

---

[4] Analytic behaviour is here intended in comparison with the numerical behaviour found previously.

For this purpose we use Figures 4.1-4 to 4.1-6. In all three cases the iteration runs several times almost symmetrically round the origin, before it decides on its broader course. In the extreme case we can convince ourselves, that there must exists a point x_g, such that the iteration can no longer escape, and indeed that each iteration only changes the sign. After two steps the original value recurs.

For x_g we must have

$$x_g = \frac{f(x_g)}{f'(x_g)} = -x_g \quad \text{or} \quad x_g = \frac{x_g^3 - x_g}{3x_g^2 - 1} = -x_g \; .$$

Simplifying the equation, we get

$$5x_g^2 = 1$$

whence the above expression.

- As regards q:

Why this quotient always has the value $q = 6$, we cannot explain here. It puzzled us too.

Further experiments make it possible to show that q is always 6 if we investigate a cubic function, whose zeros lie on the real line, and are equal distances apart. In other cases we instead find a value $q_A > 6$ and a value $q_B < 6$.

Computer Graphics Experiments and Exercises for §4.1

The only experiments for this chapter are, exceptionally, rather short. You can of course try to work out Table 4-1, or similar tables for other functions.

The next section will certainly be more interesting graphically.

4.2 Complex Is Not Complicated

In previous chapters we have formulated the two basic principles of graphical phenomena that concern us: self-similarity and boundaries. The first concept we have encountered many times, despite the different types of representation that are possible in a cartesian coordinate system.

In previous figures the boundaries between basins of attraction have not always been clearly distinguishable. In order to investigate these boundaries more carefully, we will change our previous methods of graphical representation and switch to the two-dimensional world of surfaces. We thus encounter a very ingenious and elegant style of graphics, by which we can also show the development of the boundary in two dimensions on a surface.

No one would claim that what we have discussed so far is entirely simple. But now it becomes 'complex' in a double sense. Firstly, what we are about to consider is really complicated, unpredictable and not at all easy to describe. Secondly, we will be dealing with mathematical methods that have come to be called 'calculations with complex numbers'. It is not entirely necessary to understand this type of calculation in order to generate the pictures: you can also use the specific formulas given. For that reason we will write out the important equations in full detail.   In some motivation and generalisations a knowledge of complex numbers will prove useful.   The theory of complex numbers also plays a role in physics and engineering, because of its numerous applications. This is undeniable: just look at a mathematics textbook on the subject. Independently we have collected together the basic ideas below.

The complex numbers are an extension of the set of real numbers. Recall that the real numbers comprise all positive and negative whole numbers, all fractions, and all decimal numbers. In particular all numbers that are solutions of a mathematical equation belong to this set. These might, for example, be the solutions of the quadratic equation

$$x^2 = 2,$$

namely $\sqrt{2}$, the 'square root of 2', or the famous number π, which gives the connection between the circumference and diameter of a circle. There is only one restriction. You cannot take the square root of a negative number. So an equation like

$$x^2 = -1$$

has no solutions in real numbers.   Such restrictions are very interesting to mathematicians.   New research areas always appear when you break previously rigid rules.

The idea of introducing imaginary numbers was made popular by Carl Friedrich Gauss (1777–1855), but it goes back far earlier.   In 1545 Girolamo Cardano used imaginary numbers to find solutions to the equations $x+y = 10$, $xy = 40$. Around 1550, Raphael Bombelli used them to find real roots to cubic equations.

They are imaginary in the sense that they have no position on the number line and exist only in the imagination.   The basic imaginary number is known as i and its properties are defined thus:

$$i * i = -1.$$

The problem of the equation

$$x^2 = -1$$

is thus solved at a stroke. The solutions are

$$x_1 = i \text{ and } x_2 = -i.$$

If you remember this, the rules of calculation are very straightforward.

A few examples of calculation with imaginary numbers should clarify the computational rules:[5]

- $2i * 3i = -6.$
- $\sqrt{(-16)} = \pm i*4 = \pm 4i.$
- The equation $x^4 = 1$ has four solutions 1, -1, i, and $-i$: they are all 'fourth roots' of 1.
- $6i - 2i = 4i.$

Imaginary numbers can be combined with real numbers, so that something new appears. These numbers are called *complex numbers*. Examples of complex numbers are $2+3i$ or $3.141592 - 1.4142*i$.

---

[5]You will find further examples at the end of the chapter.

A whole series of mathematical and physical procedures can be carried out especially elegantly and completely using complex numbers. Examples include damped oscillations, and the electrical behaviour of circuits that contain capacitors and resistors. In addition, deep mathematical theories (such as function theory) can be constructed using complex numbers, a fact that is not apparent when we discuss just the basics.

All equations for the basic rules, which are important for the respresentation of boundary behaviour, can be expressed using elementary mathematics.

We begin with the rule

$$i*i = -1$$

and the notation

$$z = a+i*b$$

for complex numbers.

Two numbers z_1 and z_2, which we wish to combine, are

$$z_1 = a+i*b \text{ and } z_2 = c+i*d.$$

Then the following basic rules of calculation hold:

Addition

$$z_1+z_2 = (a+i*b)+(c+i*d) = (a+c) + i*(b+d).$$

Subtraction

$$z_1-z_2 = (a+i*b)-(c+i*d) = (a-c) + i*(b-d).$$

Multiplication

$$z_1*z_2 = (a+i*b)*(c+i*d) = (a*c-b*d) + i*(a*d+b*c).$$

The *square* is a special case of multiplication:

$$z_1^2 = z_1*z_1 = (a+i*b)^2 = (a^2-b^2) + 2*i*a*b.$$

Division

Here a small problem develops: all expressions that appear must be manipulated so that only real numbers appear in the denominator. For

$$\frac{1}{z_2} = \frac{1}{c+i*d}$$

this can be achieved by multiplying by $(c-i*d)$, the *complex   conjugate* of the denominator:

$$\frac{1}{c+i*d} = \frac{c-i*d}{(c+i*d)*(c-i*d)} = \frac{c-i*d}{c^2+d^2} .$$

From this we get the rule for division:

$$\frac{z_1}{z_2} = \frac{a+i*b}{c+i*d} = \frac{a*c+b*d}{c^2+d^2} + i*\frac{b*c-a*d}{c^2+d^2} .$$

Furthermore, there is also a geometrical representation for all complex numbers and all of the mathematical operations. The two axes of an (x,y)-coordinate system are identified with the real and imaginary numbers respectively. We measure the real numbers along the x-axis and the imaginary numbers along the y-axis. Each point

P = (x,y) of this *Gaussian plane* [6] represents a complex number

$$z = x + i*y$$

(Figure 4.2-1).

Multiplication can be understood better graphically than by way of the above equation. To do this we consider not, as before, the real and imaginary parts of the complex number, but the distance of the corresponding point from the origin and the direction of the line that joins them (Figure 4.2-1). Instead of cartesian coordinates (x,y) we use *polar coordinates* (r,φ). In this polar coordinate system, multiplication of two complex numbers is carried out by the following rule:

If $z_1*z_2 = z_3$, then $r_1*r_2 = r_3$ and $\varphi_1+\varphi_2 = \varphi_3$.

We multiply the distances from the origin and add the polar angles. The distance r is described as the *modulus* of the number z and written $r = |z|$.

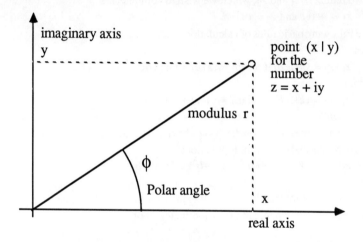

Figure 4.2-1 A point in the Gaussian plane and its polar coordinates.

What connection does the complex plane have with our mathematical experiments, with chaos, with computer graphics?

The answer is simple. Until now the region in which we have calculated, and which we have drawn, has been a section of the real–number axis, from which the parameter k was chosen. We carried out our calculations, and drew the result, for each point of that section – as long as it remained within the limits of the screen. Now we let the parameter become complex. As a result the equations for calculating chaotic systems

---

[6]*Translator's note*: Gauss represented complex numbers on a plane in about 1811. In many countries the Gaussian plane is known as the *Argand diagram*, after Jean-Robert Argand who published it in 1806. A Danish surveyor, Caspar Wessell, has a greater claim than either, having independently described the idea in 1797. The above is the conventionally recognised trio: for some reason everyone seems to ignore the fact that John Wallis used a plane to represent a complex number geometrically in his *Algebra* of 1673.

are especially simple. In particular we can represent the results directly in the complex plane. Depending on the shape of the computer screen we use a rectangular or square section of the plane. For each point in the section – as long as it remains within the limits of the screen – we carry out our calculations. The complex number corresponding to this point represents the current parameter value. After iteration we obtain the value $f(z)$ from the result of our calculation, which tells us how the corresponding screen point is coloured.

Complex numbers aren't so hard, are they? Or aren't they?

Computer Graphics Experiments and Exercises for §4.2

Exercise 4.2-1

Draw on millimetre graph paper a section of the complex plane. Using a scale 1 unit = 1 cm, draw the points which correspond to the complex numbers
$$z_1 = 2-i*2, \ z_2 = -0.5+1*1.5, \text{ and } z_3 = 2-i*4.$$
Join these points to the origin. Do the same for the points
$$z_4 = z_1+z_2 \text{ and } z_5 = z_3-z_1.$$
Do you recognise an analogy with the addition and subtraction of vectors?

Exercise 4.2-2

The following connection holds between cartesian coordinates (x,y) and polar coordinates with distance r and polar angle φ:
$$r^2 = x^2+y^2 \text{ and } \tan \varphi = y/x.$$
If $x = 0$ and $y > 0$, then $\varphi = 90°$.
If $x = 0$ and $y < 0$, then $\varphi = 270°$.
If $x = 0$ and also $y = 0$, then $r = 0$ and the angle φ is not defined.

Recall that for multiplication the following then holds: if $z_1*z_2 = z_3$ then $r_1*r_2 = r_3$ and $\varphi_1+\varphi_2 = \varphi_3$. Express this result in colloquial terms.

Investigate whether both methods of multiplication lead to the same result, using the numbers in Exercise 4.2-1.

Exercise 4.2-3

In the complex plane, what is the connection between:
- A number and its complex conjugate?
- A number and its square?
- A number and its square root?

Exercise 4.2–4

If all the previous exercise have been too easy for you, try to find a formula for powers. How can you calculate the number

$$z = (a+i*b)^p$$

when p is an arbitrary positive real number?

Exercise 4.2–5

Formulate all of the algorithms (rules of calculation) in this section in a programming language.

4.3 Carl Friedrich Gauss meets Isaac Newton

Of course, these two scientific geniuses never actually met each other. When Gauss was born, Newton was already fifty years dead. But that will not prevent us from arranging a meeting between their respective ideas and mathematical knowledge.

We transform Newton's method into a search for zeros in the complex plane. The iterative equations derived above will be applied to complex numbers instead of reals. This is a trick that has been used in innumerable mathematical, physical, and technical problems. The advantage is that many important equations can be completely solved, and the graphical representations are clearer. The normally important real solutions are considered as a special case of the complex.

Our starting point (§4.1) was

$$f(x) = x^3 - x.$$

For this the Newton method takes the form

$$x_{n+1} = x_n - (x_n^3 - x_n)/(3x_n^2 - 1).$$

For complex numbers it is very similar:

$$z_{n+1} = z_n - (z_n^3 - z_n)/(3z_n^2 - 1).$$

Recalling that $z_n = x_n + i*y_n$, this becomes

$$z_{n+1} = \frac{2*(x_n^3 - 3x_n y_n^2 + i*(3x_n^2 y_n - y_n^3))}{3x_n^2 - 3y_n^2 - 1 + i*6x_n y_n} .$$

Further calculations, in particular complex division, can be carried out more easily on a computer.

The calculation has thus become a bit more complicated. But that is not the only problem that faces us here. Now it is no longer enough to study a segment of the real line. Instead, our pictures basically lie in a section of the complex plane. This two-dimensional rectangular area must be investigated point by point. The iteration must therefore be carried out for each of 400 points in each of 300 lines.[7]

We know the mathematical result already from the previous chapter: one of the three

---

[7] These data can vary from program to program and computer to computer. Most of our pictures use a screen of 400 × 300 pixels.

zeros x_1, x_2, x_3 on the real axis will be reached.[8] This remains true even when the iteration starts with a complex number.

The graphical result is nevertheless new. To show how the three basins of attraction fit together, we have shaded them in grey in Figure 4.3-1, just as we did in §4.1.

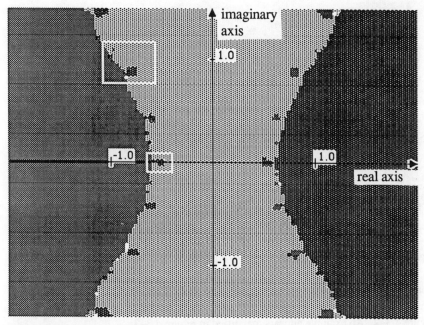

Figure 4.3-1 The basins of attraction in the complex plane.[9]

Thus the basin of x_1 is medium grey, that of x_2 is light grey, and that of x_3 is dark grey. All points for which it cannot be decided, after 15 iterations, towards which attractor they are tending, are left white.

Along the real axis, we again find the various regions that were identified in §4.1. 'Chaos', as it first appeared to us in Figure 4.1-9, we recognise in the small multi-coloured regions. We have defined chaos as the 'breakdown of predictability'. The graphical consequence of this uncertainty is fine structures 'less than one pixel in resolution'. We can investigate their form only by magnification.

The interesting region, which we investigated on the real axis in Figures 4.1-9 to 4.1-11, is shown in Figure 4.3-2 on a large scale. Again, self-similarity and regularity of the structure can be seen.

---

[8] We retain the names from §4.1, even though we are working with complex numbers z_1 etc. This is permissible, because the imaginary parts are zero.

[9] The two outlined regions on the real axis and above it will be explored in more detail in the following pictures.

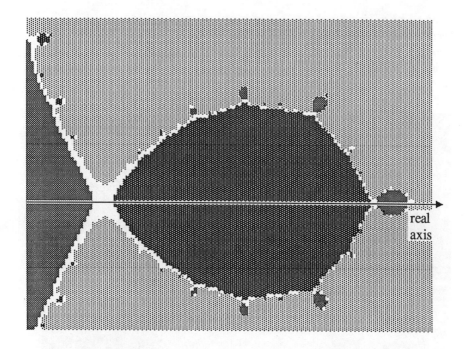

Figure 4.3-2 Section from Figure 4.3-1 left of centre.

Compared with the appearance on the real axis, which we have seen already, Figure 4.3–1 reveals something new. In many different places there appear 'grapelike' structures like Figure 4.3-2. An example appears magnified in Figure 4.3-3.

Self-similarity does not just occur on the real axis in these graphical experiments. In general, where a boundary between two basins of attraction occurs, similar figures are observed, sprinkled ever more thickly along the border. The same section as in Figure 4.3-3 leads to the next picture, in a different experiment. In this drawing only points are shown for which it cannot be decided, after 12 iterations, to which basin they belong. Thus the white areas correspond to those which in the previous pictures are shown in grey. Their structure is somewhat reminiscent of various sizes of 'blister' attached to a surface.

Further magnified sections reveal a similar scheme. The basins of attraction sprout ever smaller offshoots. One of the first mathematicians to understand and investigate such recursive structures was the Frenchman Gaston Julia. After him we call a complex boundary with self-similar elements a *Julia set*.

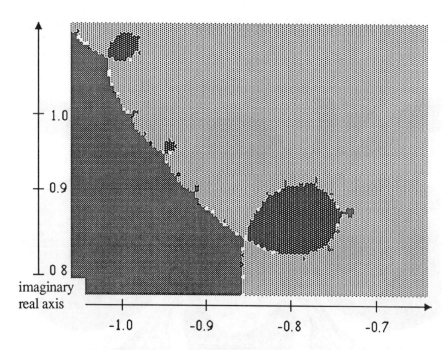

Figure 4.3-3 At the boundary between two basins of attraction.

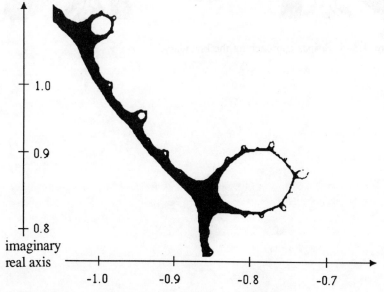

Figure 4.3-4 The boundary between two basins of attraction.

To close this section, which began with a simple cubic equation and immediately led into 'complex chaos', we illustrate a further possibility, visible in Figure 4.3-1, in graphical form. Instructions for the production of these pictures are to be found in the next chapter.

Figure 4.3-5 Stripes approaching the boundary.

5 Complex Frontiers

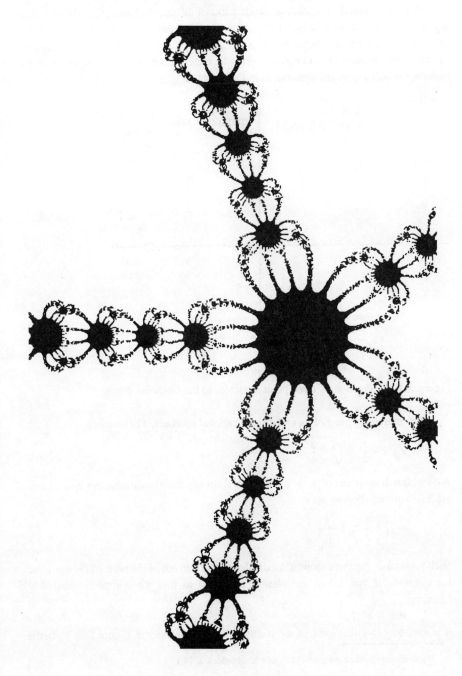

5.1   Julia and His Boundaries

We again state the question on the domains of influence of attractors. Where must we start the iteration, to be certain of reaching a given attractor? The precise boundaries between the initial zones should not be investigated. We are not exaggerating when we say that they are invisible. In order to get at least the attractors in the simplest possible fashion, we will use an arrangement as in Figure 5.1-1.

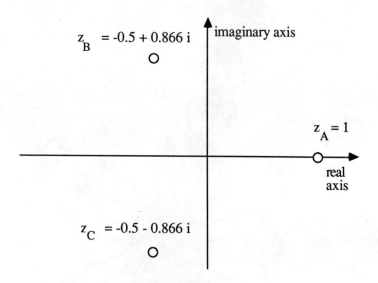

$z_B = -0.5 + 0.866\,i$

imaginary axis

$z_A = 1$

real
axis

$z_C = -0.5 - 0.866\,i$

Figure 5.1-1   Position of three point attractors in the Gaussian plane.

The imaginary parts of z_B and z_C are irrational numbers. For instance

$$z_B = -\frac{1}{2} + \sqrt{-\frac{3}{4}}\;.$$

We will finish up at one of these three points from any initial value (except $z_0 = 0$) using the following iteration equation:

$$z_{n+1} = \frac{2}{3}\,z_n + \frac{1}{3 z_n^2}\;.$$

Such points and such equations naturally don't fall from out of the blue. This one arises, for example, if you try to use Newton's method to find the complex zeros of the function[1]

$$f(z) = z^3 - 1.$$

It is easy to prove that each of the points z_A, z_B, z_C defined in Figure 5.1-1 satisfies

---

[1] If you do not see the connection here, take another look at §4.1.

$z^3 = 1$. We show this here for z_B:

$$z_B^3 = \left(-\frac{1}{2} + \sqrt{-\frac{3}{4}}\right)^3$$

$$= \left(-\frac{1}{2} + \sqrt{-\frac{3}{4}}\right) * \left(-\frac{1}{2} + \sqrt{-\frac{3}{4}}\right) * \left(-\frac{1}{2} + \sqrt{-\frac{3}{4}}\right)$$

$$= \left(-\frac{1}{2} + \sqrt{-\frac{3}{4}}\right) * \left(\frac{1}{4} - \frac{3}{4} + \sqrt{-\frac{3}{4}}\right) = \frac{1}{4} + \frac{3}{4} = 1.$$

Thus z_A, z_B, z_C are the three *complex cube roots of unity*.

We have already dealt with calculation rules for complex numbers in §4.2. If we apply Newton's method, we get

$$z_{n+1} = \frac{2}{3}(x_n + i*y_n) + \frac{x_n^2 - y_n^2 - i*(2*x_n y_n)}{3*(x_n^2 + y_n^2)^2}.$$

Thus for the complex number z_{n+1} we have

$$z_{n+1} = x_{n+1} + i*y_{n+1}$$

so that we can obtain equations for the real and imaginary parts x and y:

$$x_{n+1} = \frac{2}{3}x_n + \frac{x_n^2 - y_n^2}{3*(x_n^2 + y_n^2)^2},$$

$$y_{n+1} = \frac{2}{3}\left(y_n - \frac{x_n*y_n}{(x_n^2 + y_n^2)^2}\right).$$

In Program Fragment 5.1-1, we denote the values x_n and y_n by xN and yN, etc. The instructions for the two iteration equations can in principle be found as follows:

Program Fragment 5.1-1 (See also Program Fragments 5.1-2 and 5.1-3)

```
    ...
xN      := xNplus1;
yN      := yNplus1;
xNplus1 := 2*xN/3+(sqr(xN)-sqr(yN))
                /(3*sqr(sqr(xN)+sqr(yN)));
yNplus1 := 2*yN/3-(2*xN*yN)/(3*sqr(sqr(xN)+sqr(yN)));
    ...
```

Using these, whichever initial value $z = x + i*y$ we start with, after suitably many iterations we end up near one of the three attractors. We can recognise which by looking at the distance from the known attractors.[2] If this distance is less than some

---

[2] If we do not know the attractors, we must compare the current value z_{n+1} with the previous value z_n. If this is less than epsilon, we are finished.

preassigned bound epsilon, we say 'we have arrived'.[3]

To formulate this test in Pascal we require a boolean function. It has the value `true` if we have already reached the relevant attractor, and otherwise the value `false`. For example, for the point z_C the test becomes the following:

Program Fragment 5.1-2

```
FUNCTION belongsToZc (x, y : real) : boolean;
   CONST
      epsqu = 0.0025;
   (* coordinates of the attractor zc *)
   xc = -0.5;
   yc = -0.8660254;
BEGIN
   IF (sqr(x-xc)+sqr(y-yc) < = epsqu)
      THEN belongsToZc := true
      ELSE belongsToZc := false;
END;   (* belongsToZc *)
```

The variable `epsqu` is the square of the small number 0.05, `xc` and `yc` are the coordinates of the attractor z_C, and `x` and `y` are the working coordinates which will be modified and tested during the investigation. The calculation for the other attractors requires similar programs.

In order to obtain an overview of the basins of the attractors and the boundaries between them, we explore point by point a section of the complex plane, which contains the attractors. We colour the initial point for the iteration series according to which basin it belongs to.

We give the method for drawing the boundaries in the following Program Fragment. In a few places we have slightly modified Example 5.1-1, to make the algorithm quicker and more elegant. In particular we do not need to distinguish between x_n and x_{n+1} in the contruction of the mathematical formula. Computers work with assignments, not with formulas.

Program Fragment 5.1-3

```
PROCEDURE Mapping;
VAR
   xRange, yRange : integer;
   x, y, deltaxPerPixel, deltayPerPixel : real;
BEGIN
   deltaxPerPixel := (Right - Left) / Xscreen;
   deltayPerPixel := (Top - Bottom) / Yscreen;
```

---

[3]In fact in Program Fragment 5.1-2 we compare the square of the distance with the square of epsilon, eliminating the need to extract a square root.

```
      y := Bottom;
      FOR yRange := 0 TO Yscreen DO
      BEGIN
         x := left;
         FOR xRange := 0 to Xscreen DO
         BEGIN
            IF JuliaNewtonComputeAndTest (x, y)
               THEN SetPoint (xRange, yRange);
               x := x + deltaxPerPixel;
         END;
         y := y + deltayPerPixel;
      END;
   END; (* Mapping *)
```

In contrast to the more linear structures of the previous chapter, we no longer compute a hundred points in each series. The 120 000 points of the chosen section[4] obviously need more computing time. A complete calculation requires, in some circumstances, more than an hour. The procedure Mapping searches through the screen area one pixel at a time. For each screen point it computes the universal coordinates x and y. It passes these variables to a functional procedure, which in this case is called JuliaNewtonComputeAndTest. We use such an unequivocal name to distinguish this procedure from others which play similar roles in later programs. The corresponding screen point is coloured, or not, according to the result of this function. The procedure Mapping uses 7 global variables, which we already know from other problems:

      Left, Right, Bottom, Top,
      MaximalIteration, Xscreen, Yscreen.

For a computer with 400 × 300 pixels on the graphics screen, we might for example set up the computation as follows:

      Xscreen := 400; Yscreen := 300;
      Left := -2.0; Right := 2.0; Bottom := -1.5; Top := 1.5;

Program Fragment 5.1-4

```
      FUNCTION JuliaNewtonComputeAndTest (x, y : real) : boolean;
         VAR
            IterationNo : integer;
            finished : boolean;
            xSq, ySq, xTimesy, denominator : real;
            distanceSq, distanceFourth : real;
      BEGIN
         StartVariableInitialisation;
```

---

[4]We here refer to a section of 400 × 300 pixels.

```
REPEAT
  compute;
  test;
UNTIL (IterationNo = MaximalIteration) OR finished;
  distinguish;
END; (* JuliaNewtonComputeAndTest *)
```

Now the procedure `JuliaNewtonComputeAndTest` is formulated in reasonable generality. It makes use of four local procedures. The first sets up the values for the local variables:

Program Fragment 5.1-5

```
PROCEDURE startVariableInitialisation;
BEGIN
  finished := false;
  iterationNo := 0;
  xSq := sqr(x);
  ySq := sqr(y);
  distanceSq := xSq + ySq;
END (* startVariableInitialisation *)
```

The next procedure does the actual computation.

Program Fragment 5.1-6

```
PROCEDURE Compute;
BEGIN
  IterationNo := IterationNo + 1;
  xTimesy := x*y;
  distanceFourth  := sqr(distanceSq);
  denominator   :=
          distanceFourth+distanceFourth+distanceFourth;
  x          := 0.666666666*x + (xSq-ySq)/denominator;
  y          := 0.666666666*y -
                  (xTimesy+xTimesy)/denominator;
  xSq        := sqr(x);
  ySq        := sqr(y);
  distanceSq := xSq + ySq;
END;
```

A few tricks have been used so that the most time–consuming calculation steps, especially multiplication and division, are not carried out twice. For example the expression $\frac{2}{3}x$ is not coded as

```
2*x/3.
```

In this form, the integer numbers 2 and 3 must first be converted to real numbers every time the procedure is used. Further, a division takes more time than a multiplication. So a more efficient expression is

```
0.666666666*x.
```

After each iterative step we must test whether we have 'arrived' near enough to one of the attractors. Moreover, we must be careful that the numbers we are calculating with do not go outside the range within which the computer can operate. If that happens, we stop the calculation.

Program Fragment 5.1-7

```
PROCEDURE test;
BEGIN
    finished := (distanceSq < 1.0E-18)
        OR (distanceSq > 1.0E18)
            OR belongsToZa (x,y)
                OR belongsToZb (x,y)
                    OR belongsToZc(x,y);
END;
```

Finally we must distinguish what should be drawn[5]. The points which belong to the boundary are those which, after that maximal number of iterations, have not converged to any of the three attractors.

Program Fragment 5.1-8

```
PROCEDURE distinguish;
BEGIN
    (* does the point belong to the boundary? *)
    JuliaNewtonComputeAndTest :=
        IterationNo = MaximalIteration;
END;
```

We interpret Program Fragment 5.1-8 as follows. We include all points in the boundary for which the computation, after a given number of iterations, has not reached an attractor. In Figure 5.1-2 this `maximalIteration` = 15. In all other cases the

---

[5]We can carry out the investigation most easily when the computer has a colour graphics screen. Then each basin of attraction is given a colour, and the boundaries are easily visible.

equality condition is not fulfilled, so the iteration is stopped before the variable `iterationNo` gets that high.

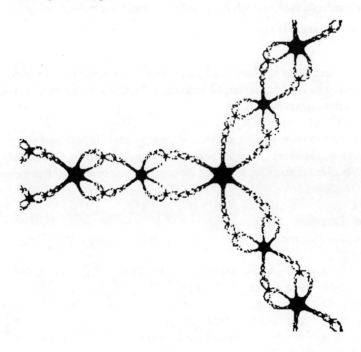

Figure 5.1-2 Boundary between the three basins after 15 iterations.

If instead of the boundary we draw the basin of attraction of one of the three attractors, as in Figure 5.1-3, we can use the functional procedure `belongsToZc` defined in Program Fragment 5.1-2.

Program Fragment 5.1-9
```
PROCEDURE distinguish
BEGIN
    (* does the point belong to the basin of zC? *)
    JuliaNewtonTestAndCompute := belongsToZc (x,y)
END;
```

It is of course entirely arbitrary, that we have chosen to draw the basin of the attractor z_C. In Exercise 5.1-2 we give hints for computing the other two basins. Perhaps you can already guess what form they will take?

Figure 5.1–3 Basin of the attractor z_C.

In a different style of drawing, as in Figure 5.1–4, we take into account the number of iterations required to reach any particular one of the three atractors. A point is then coloured if this requires an odd number of iterations. In Figure 5.1–4 you should look for the three point attractors (see Figure 5.1–1). They are surrounded by three roughly circular areas. All points in these regions of the complex plane have already reached the attractor, to the relevant degree of accuracy, after one iteration.

From there we see in turn alternating regions of black and white, from which we reach the attractor in 2, 3, 4, ... steps. In other words, the black regions correspond to initial values which take and odd number of steps to reach an attractor. In this way we obtain pictures reminiscent of *contour lines* on a map. If you think of the attractors as flat valleys and the boundaries between them as mountain ranges, this interpretation becomes even better. The peaks become ever higher, the longer the computation, deciding to which attractor the point belongs, lasts.

If the contour lines get too close together, there is always the possibility of drawing only every third or every fourth one of them.[6]

---

[6] In the Program Fragment we use the MOD function of Pascal. It gives the remainder upon division, e.g. 7 MOD 3 = 1.

Program Fragment 5.1.10

```
PROCEDURE distinguish;
BEGIN
   (* does the point reach one of the three attractors *)
   (* after an odd number of steps? *)
     JuliaNewtonComputeAndTest : =
         (iterationNo < maximalIteration)
            AND odd(iterationNo);
END;


PROCEDURE distinguish;
BEGIN
   (* does the point reach one of the three attractors *)
   (* after a number of steps divisble by 3? *)
   JuliaNewtonComputeAndTest : =
       (iterationNo < maximalIteration)
          AND (IterationNo MOD 3 = 0);
END;
```

Figure 5.1-4   Contour lines.

```
PROCEDURE distinguish;
BEGIN
    JuliaNewtonComputeAndTest : =
        (IterationNo = MaximalIteration) OR
            ((IterationNo < Bound)
                AND (IterationNo MOD 3 = 0));
END;
```

The three variations show how you can combine these methods of graphical representation to give new designs (Figure 5.1-5). The global variable Bound should be about half as big as MaximalIteration.

Figure 5.1-5 Every third contour line, and the boundaries, in a single picture.

There is a further possibility for drawing the pictures which we will only sketch here (see Execise 5.1-8). It is possible to show all three basins in the same picture. Different grey tones represent the different basins (Figure 5.1-6).

If you look at the basin of the attractor z_C (Figure 5.1-3) it may not be at all clear how the regions within it are divided up. Near the attractor itself the region is a connected piece. But at the boundaries they seem to get increasingly confused with each other. The divison nevertheless exists: it only *appears* confused. We already know from the example in Chapter 4 that the basins of attraction can never overlap.

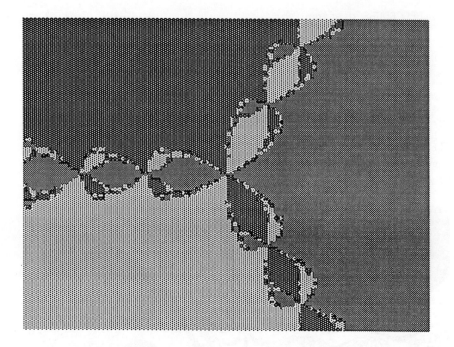

Figure 5.1-6   The basins of the three attractors z_A, z_B, z_C.

Peitgen and Richter, in Research Group in Complex Dynamics, Bremen (1984a), pp. 19, 31, describe this phenomenon by equating the three basins of attraction with the territories of three superpowers on a fantasy planet:

> 'Three power centres have divided up their sphere of influence and have agreed to avoid simple boundaries with only two adjoining regions: each boundary point must be adjacent to all three countries. If the computer graphic did not show a solution, one would scarcely believe that it existed. The key to the trick is that everywhere two countries come next to each other, the third establishes an outpost. This is then in turn surrounded by tiny enclaves of the other powers - a principle which leads to ever smaller similar structures and avoids a flat line.'

These 'similar structures' are what we have already encountered as self-similarity: here we meet them again.   Meanwhile this behaviour has become so natural to us that it is no longer a surprise!

When the basins of attraction seem so ragged, there is one thing in the picture that still hangs together: the boundary.   Here we observe a member of an entirely new class of geometrical figures, a *fractal*. This concept has been developed over the past twenty years by the Franco-Polish mathematician Benoît B. Mandelbrot.   By it he refers to structures which cannot be described by the usual forms such as lines, surfaces, or solids.

On the one hand, the boundary in Figure 5.1-2 is certainly not a surface.   For each point that we have investigated, after sufficiently many iterations, it can be assigned to one of the three attractors. Only the point (0,0), the origin, obviously belongs to the boundary. Therefore the boundary always lies between the screen pixels, and has no 'thickness'.

On the other hand, the boundary is certainly not a line. Try to work out its length! In any particular picture it looks as though this can be done. But if we magnify small pieces, we find a remarkable phenomenon: the more we magnify – that is, the more closely we look – the longer the boundary becomes. In other words, the boundary is infinitely long and has zero width. To such structures, 'between' lines and surfaces, mathematicians attribute a *fractal dimension* which is not a whole number, but lies between 1 and 2.

Two properties, closely related, are characteristic of fractals:

- *Self-similarity*, that is, in each tiny piece we observe the form of the entire shape.
- *Irregularity*, that is, there are no smooth boundaries. Lengths or areas cannot be determined.

Once this concept had been drawn to the attention of researchers, they soon found many examples of it in Nature. The coastline of an island is a fractal from the geometric point of view.

It is possible to read off the length of a coastline from a map with a given scale. But if we use a map with a different scale, that can change the result. And if we actually go to the beach, and measure round every rock, every grain of sand, every atom... we encounter the same phenomenon. The more closely we look, that is, the larger the scale of the map, the longer the coastline seems to be.

Thus many natural boundaries, by the same principle, are fractal.

Fractals with dimension between 2 and 3 are the main surfaces that will concern us. Every morning when we look in the mirror we notice that:

- Skin is fractal – especially when one is 'getting on in years'. It has the basic purpose of covering the body with the smallest possible surface. But for several other reasons a crumpled structure is preferable.

It is even more obvious, when we leave the house:

- Clouds, trees, landscapes, and many other objects, when viewed at a different level, appear fractal. To investigate these is the subject of rapidly growing research activity in all areas of natural science.

The consequences of this discovery for biology are explained in Walgate (1985) p.76. For small creatures, for example insects on a plant, the living space grows in an unsuspected fashion. Consider as a model case two species, differing in size by a factor of 10. They must expand their population on one of the available surfaces. There should therefore be 100 times as many small creatures as large. But this argument takes on a different appearance if we think of the surface of a leaf from the

Figure 5.1-7   A Julia set with fivefold symmetry.

fractal point of view.   Then we find about 300 to 1000 times as many small insects in the same space that a single creature, 10 times as big, would occupy. This has actually been verified experimentally: see Morse *et al.* (1985). Thus there are many more tiny insects on a plant than had hitherto been thought. 'The smaller the organisms, the larger the world in which they live.'

An entire series of physical processes, whose detailed description raises many other problems, are of a fractal nature.   Examples include Brownian motion and the study of turbulence in the flow of gases and fluids. This is true even though natural fractals can obviously display self-similarity and piecewise crumpling only up to some limit, whereas mathematical fractals have these properties completely and at every level.

As often happens in mathematics, parts of the scientific preparatory work had already been carried out some time ago. But the discoveries of the French mathematicans Pierre Fatou and Gaston Julia were already becoming somewhat forgotten. In honour of the French researcher who studied the iteration of complex functions prior to 1920, fractal boundaries (Figure 5.1-2) are known as *Julia sets*.

The possibilities raised by the computer now make it possible, for the first time, to investigate this fundamental area. It is indisputably to the credit of the Bremen Research Group of H. O. Peitgen and P. Richter to have excited attention - not only in the technical world - with the results of the 'computer cookery' of their work in graphics.

Their pioneering work is well known internationally.

The investigation of Newton's method carried out above, for the equation
$$z^3 - 1 = 0,$$
can be applied equally well to other functions.   To show you one result, Figure 5.1-7 indicates what happens for
$$z^5 - 1 = 0.$$
You will find some hints about it in Exercises 5.1-4 and 5.1-5.

In other exercises for this chapter we suggest experiments, leading to innumerable new forms and figures, which we have scarcely been able to explore ourselves.   So varied are the possibilities opened up, that you are guaranteed to produce pictures that nobody has ever seen before!

Computer Graphics Experiments and Exercises for §5.1

Exercise   5.1-1

Apply Newton's method to the function
$$f(z) = z^3 - 1.$$
In the iteration equation
$$z_{n+1} = z_n - \frac{f(z_n)}{f'(z_n)}$$
insert the expressions for the functions $f(z)$ and $f'(z)$.   Show that this leads to the equation
$$z_{n+1} = \frac{2}{3} z_n + \frac{1}{3z_n^2} .$$

For the complex numbers z_A to z_C in Figure 5.1-1 compute the value of z^3. What happens?

Exercise   5.1-2

On the basis of Program Fragments 5.1-1 to 5.1-3, write a program that allows you to compute the basins and boundaries of the three attractors in Figure 5.1-3.

Next investigate and draw the basin of the attractor
$$z_A = 1$$
with the iteration formula
$$z_{n+1} = \frac{2}{3} z_n + \frac{1}{3z_n^2} .$$

Compare the resulting picture with those that you get for

$$z_B = -0.5 + 0.8660254 * i$$

and

$$z_C = -0.5 - 0.8660254 * i,$$

see Figure 5.1-3.

It is not a coincidence that the pictures resemble each other in various ways. The reason involves the equivalence of the three complex cube roots of unity.

Exercise 5.1-3

Using the method of magnifying detail, investigate sections of Figures 5.1-1 to 5.1-6. Choose regions that lie near boundaries. In all these cases we find self-similarity!

Check that if we increase the number of iterations (and also increase the admittedly not short computation time) self-similar structures continue to appear.

Exercise 5.1-4

If you want to apply Newton's method with powers higher than 3, for example to equations such as $z^4-1 = 0$ or $z^5-1 = 0$, you must know the appropriate complex roots of 1. These give the positions of the attractors.

In general the nth roots of 1 are given by

$$z_k = \cos\left(\frac{k*360°}{n}\right) + i*\sin\left(\frac{k*360°}{n}\right)$$

where k runs from 0 to $n-1$. Produce (with the aid of a computer program) a table of roots of unity z_n up to $n = 8$.

Exercise 5.1-5

Investigate and draw Julia sets resulting from Newton's method for
$$z^4-1 = 0 \quad \text{and} \quad z^5-1 = 0.$$
The mathematical difficulties are not insuperable. In Program Fragment 5.1-3 you must construct a modified function `JuliaNewtonComputeAndTest`. If you cannot solve this and the following exercises straight away, you can take a look at §6.4. There we summarise the important rules for calculating with complex numbers.

Exercise 5.1-6

We apply Newton's method and thereby get beautiful, symmetric computer graphics. Although the above examples seem to be well founded, they are certainly not to be thought of as 'sacred'. As with our excellent experience in changing formulas in the previous chapter, so too we can change Newton's method somewhat.

Starting from z^p-1, we insert in the equation

$$z_{n+1} = z_n - \frac{f(z_n)}{f'(z_n)}$$

the factor v (which is of course complex), obtaining:

$$z_{n+1} = z_n - v*\frac{f(z_n)}{f'(z_n)} \ .$$

First decide for yourself, without drawing, what the attractors will be. How far do they correspond to the complex roots of unity, that is, with the solutions to $z^p - 1 = 0$?

Start with values v that lie near 1. Investigate the influence of v on the form of the Julia sets.

Exercise 5.1-7

Investigate the changes that occur if we modify the formula by putting an imaginary summand i*w in the denominator:

$$z_{n+1} = z_n - \frac{f(z_n)}{i*w+f'(z_n)} \ .$$

Again work out, without drawing, what the attractors are. Draw the way the basins of attraction fit together.

How far does the modification of the equation influence the symmetry of the corresponding pictures?

Exercise 5.1-8

If you wish to work with grey tones, as in Figure 5.1-6, but these are not directly available on your computer, Figure 5.1-8 shows how to proceed.

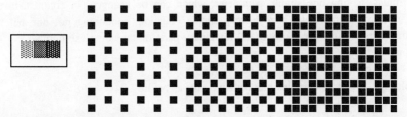

Figure 5.1-8   Grey shades, normal size (left) and magnified, to show individual pixels.

To achieve this, not every point should be coloured, even when after calculation it is known in which basin it belongs. The drawing depends in which basin and also in which row or column of the screen a point occurs.

In particular the conditions for drawing points should be:

```
IF (odd(row) AND (column MOD 4 = 0))        (* light grey *)
   OR (NOT odd (row) AND (column MOD 4 = 2)) THEN ...

IF (odd(row) AND odd(column))               (* medium grey *)
   OR (NOT odd (row) AND NOT odd(column)) THEN ...

IF (odd(row) AND (column MOD 4 <> 0))       (* dark grey *)
   OR (NOT odd (row) AND (column MOD 4 <> 2)) THEN ...  .
```

5.2 Simple Formulas give Interesting Boundaries

The various pictures of previous pages, and the innumerable sections and variations that you have derived from them, owe their existence to the calculating performance of the computer. Without this, no one would have made the effort to perform hundreds of thousands of completely different calculations – especially since it appears that the pictures become ever more ragged and complicated, the more complicated (and ragged?) the underlying formulas become.

Again, this conjecture was first considered by B. B. Mandelbrot,[7] who already knew in 1979–80 that the production of 'richly structured' pictures does not necessarily need complicated mathematical formulas. The important thing is for the iteration formula to be nonlinear. It can thus contain polynomials of the second or higher degree, or transcendental functions, or other such things.

The simplest nonlinear iteration equation that leads to nontrivial results was suggested by Mandelbrot:

$$z_{n+1} = z_n{}^2 - c.$$

That is, we obtain a new member of the iteration series, by taking the previous one, squaring it, and subtracting from the square a number c.

Until now, in previous chapters, we have investigated relatively complicated formulas, having one simple parameter, which can be changed. In contrast, in this Mandelbrot formula we have a very simple equation, containing not one but two parameters. These are the real and imaginary parts of the complex[8] number c. The equation

$$c = c_{real} + i * c_{imaginary}$$

holds for c. For the complex number z we have as before

$$z = x + i * y.$$

In other words, we can write the formula as

$$z_{n+1} = x_{n+1} + i * y_{n+1}$$
$$= f(z_n)$$
$$= (x_n{}^2 - y_n{}^2 - c_{real}) + i * (2 * x_n * y_n - c_{imaginary}).$$

---

[7]See the very readable article *Fractals and the Rebirth of Iteration Theory* in Peitgen and Richter (1986) p. 151.

[8]If the mathematical concepts of real, imaginary, and complex numbers still cause difficulty, we again advise you to reread §4.2

These formulas can be set up within a Pascal program:

Program Fragment 5.2-1

```
...
xSq    := sqr(x)
ySq    := sqr(y)
y      := 2*x*y - cImaginary
x      := xSq - ySq - cReal


...
```

It is important to keep these statements in the given order, or else information will get lost.[9]   In particular, contrary to alphabetical order and other custom, we must first compute the value of y and then that for x (because x appears in the equation for y).

Just like those for Newton's method, these iteration formulas seem to be fairly harmless and not very complicated.   But it is not so easy to grasp the possibilities for investigation that arise when a single parameter is changed.   Not only the components of c play a role, but also the initial value of z, the complex number

$$z_0 = x_0 + i * y_0.$$

Thus we have four quantities to change and/or draw, namely

$$x_0, y_0, c_{real}, c_{imaginary}.$$

On a sheet of paper we can show only two dimensions, so we must choose two out of the four as the basis of our drawing.   As in the previous chapter these are the components x_0 and y_0 of the complex initial value z_0.   The computation of these pictures takes the following form.   The position of a point on the screen (in screen coordinates) represents the components x_0 and y_0 (in universal coordinates).   For a given value

$$c = c_{real} + i * c_{imaginary}$$

the iteration is carried out.   As a result, x and y change, and with them z.   After a given number of iterations we colour the point corresponding to z according to the result.   If it is not possible to use different colours, we can still use black–and–white or grey tones.   The method is then repeated for the next value of z_0.

The Mandelbrot formula is so constituted that it has only two attractors.   One of them is 'infinite'.   By the attractor '∞', we mean that the series of numbers $f(z)$ exceeds any chosen value.   Since for complex numbers there is no meaningful concept of larger/smaller, we understand by this that the square of the modulus, $|f(z)|^2$, exceeds any given value after sufficiently many steps.   It does not matter much which value, and we take for this bound the number 100.0 (defined as a real number so that the comparison does not take too long).

The variable MaximalIteration can be seen as a measure of how much patience we have.   The bigger this number, the longer the calculation takes, and the longer we must wait for the picture.   But if MaximalIteration is too small, the drawing is very

---

[9]You will still get interesting pictures - but not the ones intended.

poor.   If $|f(z)|^2$ stays below the value 100.0 after this number of steps, we declare the attractor to be 'finite' or effectively 'zero'.

In fact the situation is somewhat more complicated. Only in some cases is $f(z) = 0$ a limiting value. In other cases we encounter a different number $z \neq 0$, a so-called *fixed point.* It lies near the origin, and satisfies $f(z) = z$.   Further, there are also attractors that are not single points, but which consist of 2, 3, 4,... or more points. For an attractor with the period 3 we have

$f(f(f(z))) = z$.

Despite these complications, we concentrate on only one important thing: that the attractor be finite, that is, that the series should not exceed the value 100.0.

Each picture shows a section of the complex z-plane. We make the initial values z_0 the basis of our drawing.   At each iteration z changes, and we give it the value computed for $f(z)$ at the previous stage.

The complex parameter c must be kept constant throughout a given picture.

To repeat: there exist two attractors, zero and infinity, whose basins of attraction border each other. As has already been explained in §5.1, we can call these complex boundaries Julia sets.   For the rest of this chapter we will be concerned with their graphical representation.

Two different complex numbers c_1 and c_2 generate two different Julia sets, and thus two different graphics!   The variety of possible complex numbers produces an uncountable number of distinct pictures.

The graphical appearance of these sets can be very varied too, because the form of the Julia set depends strongly on the value of the parameter c.

For some of the pictures on the following two pages only the upper halves are drawn. They are symmetric about the origin, as can be seen in Figure 5.2-4 (lower left) and Figure 5.2-5 (upper right). In the series Figures 5.2-1 to 5.2-8, the complex parameter c takes the values

$c_1 = 0.1 + 0.1*i, c_2 = 0.2 + 0.2*i, ... , c_8 = 0.8 + 0.8*i$.

We start with $c = 0$, surely the simplest case. Without a computer it is easy to see that the boundary is the unit circle. Each z-value, whose modulus is greater than 1, has ∞ as an attractor. Each value $|z| < 1$ has 0 as attractor and should be drawn. We also colour the points with $|z| = 1$ since these too lead to a finite attractor.   In terms of contours nothing much can be distinguished in this case, and we do not give a picture. But in the majority of cases, when we change c, the resulting picture yields contours. If c is increased in steps of $0.1 + 0.1 * i$, we get in turn Figures 5.2-1 to 5.2-8.

A program to generate these figures obviously has certain parts in common with the graphics programs that we have already encountered (see Program Fragment 5.2-2). The remainder can be found in Program Fragment 5.1-3.   Make the procedure now called JuliaComputeAndTest call the procedure Mapping. The surrounding program must supply the global variables cReal and cImaginary with suitable values.

In comparison with Program Fragment 5.2-1 a small improvement has been made.

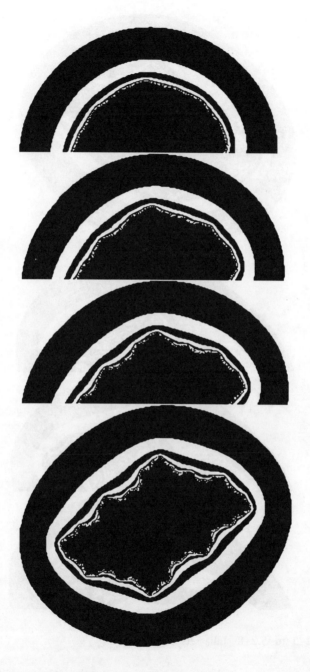

Figures 5.2-1 to 5.2-4 Julia sets.

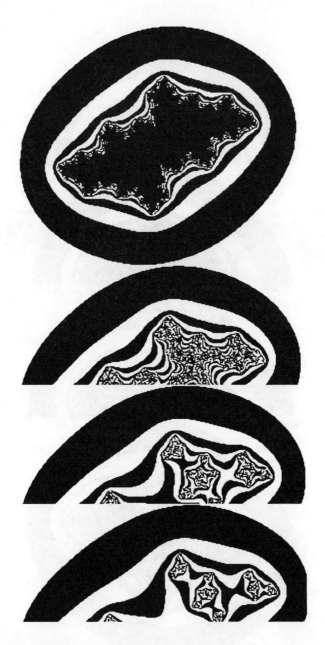

Figures 5.2–5 to 5.2–8 Julia sets.

As a result we save one multiplication per iteration and replace it by an addition, which is computed significantly faster.

Below we describe a complete functional procedure `JuliaComputeAndTest`, containing all relevant local functions and procedures.

Program Fragment 5.2-2

```
FUNCTION JuliaComputeAndTest (x, y : real) : boolean;
  VAR
      iterationNo   : integer;
      xSq, ySq, distanceSq   : real;
      finished : boolean;
  PROCEDURE startVariableInitialisation;
  BEGIN
      finished := false;
      iterationNo := 0;
      xSq := sqr(x); ySq := sqr(y);
      distanceSq := xSq + ySq;
  END;  (* startVariableInitialisation *)

  PROCEDURE compute;
  BEGIN
      iterationNo := iterationNo  + 1;
      y := x * y;
      y : = y + y - cImaginary;
      x := xSq - ySq - cReal;
      xSq := sqr(x); ysQ := sqr(y);
      distanceSq := xSq + ySq;
  END;  (* compute *)

  PROCEDURE test;
  BEGIN
      finished := (distanceSq > bound);
  END;  (* test *)

  PROCEDURE distinguish;
  BEGIN  (* does the point belong to the Julia set? *)
      JuliaComputeAndTest :=
          iterationNo = maximalIteration;
  END;  (* distinguish *)

BEGIN (* JuliaComputeAndTest *)
  startVariableInitialisation;
  REPEAT
```

```
    compute;
    test;
UNTIL (iterationNo = maximalIteration) OR finished;
    distinguish;
END; (* JuliaComputeAndTest *)
```

We recognise familiar structures in the Program Fragment. In the main part the screen section is scanned point by point and the values of x and y passed to the functional procedure `JuliaComputeAndTest`. These numbers are the starting values x_0 and y_0 for the iteration series. The global constants `cReal` and `cImaginary` control the form of the set that is drawn.

Each new pair of numbers generates a new picture!

For mathematicians, it is an interesting question, whether the Julia set is connected. Can we reach every point in the basin of the finite attractor, without crossing the basin of the attractor ∞? The question of connectedness has been answered in a difficult mathematical proof, but it can be studied rather more easily with the aid of computer graphics.

In Figures 5.2-7 and 5.2-8 we certainly do not have connected Julia sets. The basin of the attractor ∞ can be seen from the contour lines. It cuts the Julia set into many pieces. In Figures 5.2-1 to 5.2-5 the Julia set is connected. Is this also the case in Figure 5.2-6? We ask you to consider this question in Exercise 5.2-1.

As another example we will demonstrate what effect an extremely small change of c can have on the picture of the Julia set. We choose for the two parameters c_1 and c_2 the following values, which differ only by a very tiny amount:

$$c_1 = 0.745\ 405\ 4 + i*0.113\ 006\ 3$$
$$c_2 = 0.745\ 428\ 0 + i*0.113\ 009\ 0$$

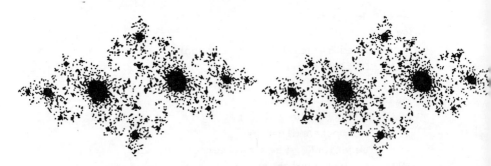

Figures 5.2-9 and **5.2-10** Julia sets for c_1 and c_2.

In both cases the Julia sets appear the same (Figures 5.2-9, 5.2-10). The pictures that follow are drawn with contour lines, so that the basin of the attractor ∞ can be deduced from the stripes. The object of investigation is the indicated spiral below and to the right of the middle of Figures 5.2-9 and 5.2-10.   Figures 5.2-11 (the same as Figure 5.2-9 magnified 14 times) and 5.2-12 (magnified about 135 times) are here drawn only for c_1.   These magnified pictures are also barely distinguishable from those for c_2.

Figure 5.2-11 Julia set for c_1. Section from Figure 5.2-9.

Figure 5.2-12 Julia set for c_1. Section from Figure 5.2-11.

In Figure 5.2-13 (for c_1) and 5.2-14 (for c_2) the pictures of the two Julia sets first begin to differ in detail in the middle. After roughly 1200-fold magnification there is no visible difference at the edge of the picture, at least up to small displacements.

Figure 5.2-13 Julia set for c_1. Section from Figure 5.2-12.

Figure 5.2-14 Julia set for c_2. Section corresponding to Figure 5.2-13.

The extreme 6000-fold magnification in Figure 5.2-15 (for c_1) and Figure 5.2-16 (for c_2) confirms the distinction.

The stripes indicating the basin of attraction of ∞ are connected together in Figure 5.2-15 from top to bottom. At this point the figure is divided into a left and a right half. As a result the Julia set is no longer connected. It is different for c_2 in Figure 5.2-16. The basin of attraction of ∞ 'silts up' in ever more tiny branches, which do not touch each other. Between them the other attractor holds its ground.

We must magnify the original Figures 5.2-9 and 5.2-10, with an area of about 60 cm^2, so much that the entire figure would cover a medium-sized farm (21 hectares). Only then can we notice the difference.

Figure 5.2–15 Julia set for c_1. Section from Fig 5.2-13.

Figure 5.2–16 Julia set for c_2. Section from Fig 5.2-14.

Who would have realised, just a few years ago, that mathematicians might use computer graphics to investigate and test basic mathematical concepts? This trend has led to a new research area for mathematicians - *experimental mathematics*. Basically they now work with similar methods to those used long ago by physicists. But the typical measuring instrument is not a voltmeter, but a computer.

We have no wish to give the impression that the only interest in Julia sets is for measurement or research. Many pictures are also interesting for their aesthetic appeal, because of the unusual shapes that occur in them.

Throughout this book you will find pictures of Julia sets, and we recommend all readers to work on them with their own computers.   Each new number pair

c_{real}, $c_{imaginary}$ produces new pictures.   To be able to carry out the investigation in a fairly systematic way, we will first look at the sets that can be generated using c_1 and c_2.

The following pictures are generated using the same data as in Figure 5.2-10. They differ from it mostly in their size, and in the way that the ideas under investigation are represented.

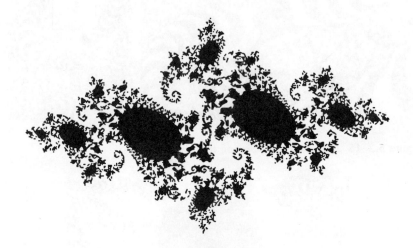

Figure 5.2-17 Julia sets with small iteration number.

As we have already shown earlier in this chapter, the form of this figure is completely connected.   But that does not mean that every pixel, for which we compute the iteration series, really belongs to the figure.   Because of the filigreed fractal structure of many Julia sets it is possible that in a given region all points which we investigate just lie near the figure.   This is how the apparent gaps in Figure 5.2-17 arise.   This holds even more when we raise the iteration number.   Then there is often just 'dust' left.   In Figure 5.2-18 we have therefore illustrated another limitation.   It shows the points for which it is already clear after 12 iterations that they do not belong to the Julia set.

Even though these stripes, which we have referred to as 'contour lines', do not count towards the Julia set, they represent an optical aid without which the fine fractal structure could not often be detected.

If these stripes do not seem as thick and dominant as in the previous pictures, this is because not every second, but every third of them has been drawn.   We interpret these regions as follows.   The outermost black stripe contains all points, for which it is already apparent after 3 iterations steps that the iteration sequence converges to the attractor ∞. The next stripe inwards represents the same for 6 iterations, and so on.

It is also apparent that we cannot show all details of a fractal figure in a single picture.   With a relatively coarse solution Figure 5.2-17 shows many different spiral shapes.   On looking closer they disappear, to be replaced by finer structures inside the

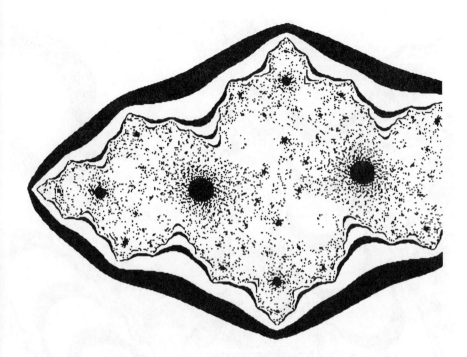

Figure 5.2-18 Julia set with higher iteration number also acts as a boundary.

large black regions.

If we investigate further details of this figure, you will perhaps notice the contiguous spirals in the middle.   On the next two pages we show successive magnifications of this motif. Pay special attention to the central point. That is the origin of the coordinate system.

There we see an already known phenomenon (§2.2) in a new guise: a period-doubling scenario!   The two contiguous spirals generate several smaller spirals, which you can see at the centre of Figure 5.2-19.   After even greater magnification this can be seen in Figure 5.2-20, when it is possible to detect sixteen diminutive mini-spirals.

What else do you think is concealed within the black central dot?

Near the filigreed shape of the Julia set you can see still more 'contour lines' which, together with the white areas, belong to the basin of the attractor ∞.

Note also the numerous details of the Julia set in these pictures. In particular, as will become clearer, it looks as though all structures eventually belong to contiguous spirals.

At the beginning of this chapter, we stated that each c-value produces a different picture. Now we have seen that it can even be more than one picture.

Not all c-values lead to figures as rich in detail as the value $c \sim 0.745+0.113*i$,

Figure 5.2-19   Section from the centre of Figure 5.2-17.

Figure 5.2-20   Section from the centre of Figure 5.2-19.

which underlies Figures 5.2-9 to 5.2-20. If you try out other values yourself, it can in some circumstances be rather boring, sitting in front of the computer, watching a picture form. For this reason we will suggest another way to draw the pictures, which is suitable for a general overview. To avoid confusion, we refer to the previous method as Mapping, while the new one is called backwardsIteration.

We begin with the following observation.

Each point that lies outside the Julia set follows, under the iteration

$$z_{n+1} = z_n^2 - c,$$

an ever larger path. It wanders 'towards infinity'. But we can reverse the direction. Through the backwards iteration

$$z_n = \sqrt{(z_{n+1} + c)}$$

the opposite happens: we get closer and closer to the boundary. Starting from a very large value such as $z = 10^6 + 10^6 i$ we reach the boundary after about 20 to 30 steps. Because each root has two values in the field of complex numbers, we get about 2^{20} to 2^{30} points ($\sim 10^6$ to $\sim 10^9$), which we can draw. This high number of possible points also makes it possible to stop the calculation at will.

In a Pascal program we implement this idea with a procedure backwards, which calls itself twice. This style of recursive programming leads to rather elegant programs.

The procedure backwards requires three parameters: the real and imaginary components x and y of a point, as well as a limit on the number of recursive steps. Inside the procedure the roots of the complex number are taken and the result is stored in two local variables xLocal and yLocal. Once the limit on the depth of recursion is reached, the two points corresponding to the roots are drawn. Otherwise the calculation continues. The constant c is added to the roots, and a new incarnation of the procedure backwards is called.

Extracting roots is very easy for complex numbers. To do so we go over to the polar coordinate representation of a number. As you know from §4.2, r and φ can be computed from x and y. Recalling the rule for multiplication, we see that the square root of a complex number is obtained by halving the polar angle φ and taking the (usual) square root of the radius r.

Program Fragment 5.2-3

```
PROCEDURE backwards (x, y : real; depth : integer);
  VAR
      xLocal, yLocal : real;
  BEGIN
     compRoot (x, y, xLocal, yLocal);
     IF depth = maximalIteration THEN
     BEGIN
        SetUniversalPoint (xLocal, yLocal);
        SetUniversalPoint (-xLocal, -yLocal);
     END
```

```
      ELSE IF NOT button THEN (*button: break calculation *)
      BEGIN
         backwards (xLocal+cReal, yLocal+cImaginary, depth+1);
         backwards (-xLocal+cReal, -yLocal+cImaginary,
                         depth+1);
      END;
   END  (* backwards *)
```

Program Fragment 5.2-4

```
   PROCEDURE compRoot (x, y: real; VAR a, b : real);
      CONST
         halfPi = 1.570796327;
      VAR
         phi, r : real;
   BEGIN
      r := sqrt(sqrt(x*x+y*y));
      IF ABS(x) < 1.0E-9 THEN
         BEGIN
            IF y > 0.0 THEN phi := halfPi
               ELSE phi := halfPi + pi;
         END
      ELSE
         BEGIN
            IF x > 0.0 THEN phi := arctan (y/x)
               ELSE phi := arctan (y/x) + pi;
         END;
      IF phi < 0.0 THEN phi := phi + 2.0*pi;
      phi := phi*0.5;
      a := r*cos(phi); b := r*sin(phi);
   END;  (* compRoot *)
```

If you want to experiment with this version of the program, you must set the values of the real part cReal and the imaginary part cImaginary of the complex parameter c. Take a maximal iteration depth of, e.g.,

```
      maximalIteration := 30;
```

If you make this number too large you will get a *stack overflow error*. For each new recursion step the computer sets aside yet another storage location.

In preparation, call the procedure with

```
      backwards (1000,1000,1);
```

You can quit the procedure by pressing a key or clicking the mouse, whichever is set up on your computer (see chapter 11). You must hold the key down long enough so that

Figure 5.2-21 Backwards iteration, 20 seconds' computing time.

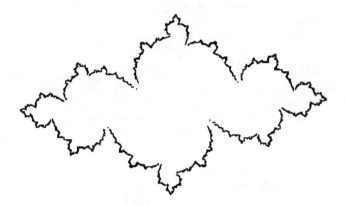

Figure 5.2-22 Backwards iteration, 4 hours' computing time.

you exit from the entire set of incarnations of the recursive procedure.

The next two pictures show examples of this method. In Figure 5.2-21 you see that already after a few seconds the general from of the Julia set is visible. The applicability of this fast method is limited by its duration. The points that lie far in from the edge are scarcely reached. Even after a few hours of computing time (Figure 5.2-22) the interesting spiral structures, which we really should expect, cannot be seen.

Computer Graphics Experiments and Exercises for §5.2

Exercise 5.2-1

Write a program to compute and draw Julia sets. It should have the facility for cutting out sections and magnifying them.

Preferably the computation should follow the 'contour line' method.

Do not forget to document your programs, so that you can pursue other interesting questions later.

Use the program to investigate the first magnification of Figure 5.2-6, to decide the question: is the Julia set connected or does it fall to pieces?

Exercise 5.2-2

Investigate similar series as in Figures 5.2-1 to 5.2-8.

The c-values can be changed along the real or imaginary axis. Can you detect – despite the complexity – some system in these series?

To save time you can exploit the symmetry of these Julia sets. You need carry out the computation only for half the points. The other half of the picture can be drawn at the same time or with the aid of some other graphical program.

Exercise 5.2-3

Find particularly interesting (which can mean particularly wild) regions, which you can investigate by further magnification.

Note the corresponding c-values, and try out similar values.

How do the pictures corresponding to two conjugate c-values differ from each other?

(If $c = a + i*b$ then its complex conjugate is $a - i*b$.)

Exercise 5.2-4

Build the procedure for backwards iteration into your program.

Use it to investigate a larger number of parameters c.

- Is the boundary of the set smooth or ragged?
- Does the set appear to be connected or does it split into several pieces?

Exercise 5.2-5

What pictures arise if, instead of large values, you start with small ones? For example,

```
backwards (0.01, 0.01, 1).
```
Compare the pictures with those in Exercise 5.2-1.

Can you give an explanation?

Exercise   5.2-6

Write a program which produces a series of Julia sets, so that the c-values are either

* random

or

* change step by step in a predetermined fashion.

Exercise   5.2-7

Attach a super-8 camera or a video camera which can take single frames to your computer.   Make a film in which you compute several hundred members of a sequence such as Figures 5.2-1 to 5.2-8, to show how the shape of the Julia set changes as the value of c is altered.   The results make the world of Julia sets seem rather more orderly that one might expect from the individually computed pictures.

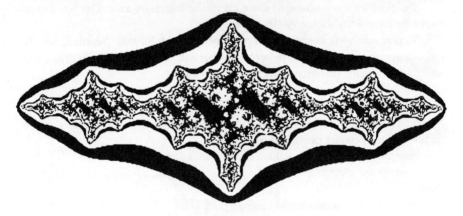

Figure 5.2-23 Yet another Julia set (just to whet your appetite).

6 Encounter with the Gingerbread Man

Pictures like grains of sand on the seashore... the graphics which we can generate by the methods of the previous chapter are as different or as similar as sand. Each complex number produces another picture, sometimes fundamentally different from the others, sometimes differing only in details. Despite the self-similarity (or maybe because of it?) new surprises appear upon magnification.

6.1 A Superstar with Frills

From all this variety of apparent forms we will pursue only one property. It concerns the question: is the picture connected or does it split apart?[1] But instead of investigating the connectivity with particular magnifications in unpredictable places, we wihall use a trick. Perhaps something has already occurred to you in connection with the experiments in Chapter 5?

As an example, consider the two complex numbers
$$c_1 = 0.745\ 405\ 4 + i*0.113\ 006\ 3$$
and
$$c_2 = 0.745\ 428 + i*0.113\ 009.$$
We have already looked at the corresponding pictures in Chapter 5 (Figures 5.2-9 to 5.2-16), and discovered that the Julia set corresponding to c_2 is connected. In contrast c_1 produces pictures in which the basin of the finite attractor splits into arbitrarily many pieces. It suffices to show that the figure is not connected together at one place, but falls into two pieces there. The more general conclusion follows by self-similarity.

Consider Figures 6.1-1 and 6.1-2, corresponding to the above c-values. They show at about 40-fold magnification the region in the neighbourhood of the origin of the complex plane, which lies symmetrically placed in the middle of the entire figure.

The question whether the set is connected can now be answered quite easily. If you look in Figure 6.1-1 in the middle of the striped basin, it is clear that no connection can exist between the lower left and the upper right portions of the figure. That is a fundamentally different observation from Figure 6.1-2. There the stripes in the basin of the attractor at infinity do not approach near enough to each other.

In the middle there appears, between them, a relatively large region from the other attractor. We have already encountered something similar in Figures 5.2-14 and 5.2-16. You can show for yourself that such circle-like forms appear at very many places in this Julia set.   But this region round the origin of the complex plane is fairly large. Thus we can pursue the investigation without any pictures at all, reducing the question to a single point, namely the origin. In the sequence for c_1 (Figure 6.1-1) we have drawn the required boundary after about 160 iterations, and thereby made sure that the origin belongs to the basin of the attractor ∞. But even if the computer takes a week, we cannot determine the boundary for c_2 (Figure 6.1-1).

---

[1]We call a set *connected* if there is a path within it, by means of which we can reach any point without leaving the set. If this is not the case, the set must divide into two distinct parts.

Figure 6.1-1 Julia set for c_1, section near the origin.

Figure 6.1-2 Julia set for c_2, section near the origin.

In other words, the Julia set for a particular number c in the iteration sequence

$$z_{n+1} = z_n^2 - c$$

is connected, provided the sequence starting from

$$z_0 = 0$$

does not diverge.

We shall not pursue the mathematics of this relationship any further. All iteration sequences depend only on c, since $z_0 = 0$ is predetermined. If you insert this value into the iteration sequence, you obtain in turn:

$$z_0 = 0,$$
$$z_1 = -c,$$
$$z_2 = c^2 - c,$$
$$z_3 = c^4 - 2c^3 + c^2 - c,$$
$$z_4 = c^8 - 4c^7 - 2c^6 - 6c^5 + 5c^4 - 2c^3 + c^2 - c,$$

etc.

Whether this sequence diverges depends upon whether the positive and negative summands are of similar sizes, or not. Then, when the modulus of z exceeds a certain bound, squaring produces such a large increase that subtracting c no longer leads to small numbers. In the succeeding sequence the z-values grow without limit.

				Quadratic Iteration / c real					
1	**2**	**3**	**4**	**5**	**6**	**7**	**8**	**9**	**10**
c →	0	1,00	-1,00	-0,50	-0,25	1,50	2,00	2,10	1,99
n ↓	Z n	Z n	Z n	Z n	Z n	Z n	Z n	Z n	Z n
0	0,00	0,00	0,00	0,00	0,00	0,00	0,00	0,00	0,00
1	0,00	-1,00	1,00	0,50	0,25	-1,50	-2,00	-2,10	-1,99
2	0,00	0,00	2,00	0,75	0,31	0,75	2,00	2,31	1,97
3	0,00	-1,00	5,00	1,06	0,35	-0,94	2,00	3,24	1,89
4	0,00	0,00	26,00	1,63	0,37	-0,62	2,00	8,37	1,59
5	0,00	-1,00	677,00	3,15	0,39	-1,11	2,00	68,00	0,53
6	0,00	0,00	*****	10,44	0,40	-0,26	2,00	*****	-1,71
7	0,00	-1,00	*****	109,57	0,41	-1,43	2,00	*****	0,94
8	0,00	0,00	*****	*****	0,42	0,55	2,00	*****	-1,11
9	0,00	-1,00	*****	*****	0,42	-1,19	2,00	*****	-0,75
10	0,00	0,00	*****	*****	0,43	-0,08	2,00	*****	-1,43
45	0,00	-1,00	#NUM!	#NUM!	0,48	-1,45	2,00	#NUM!	-1,98
46	0,00	0,00	#NUM!	#NUM!	0,48	0,60	2,00	#NUM!	1,92
47	0,00	-1,00	#NUM!	#NUM!	0,48	-1,13	2,00	#NUM!	1,68
48	0,00	0,00	#NUM!	#NUM!	0,48	-0,21	2,00	#NUM!	0,84
49	0,00	-1,00	#NUM!	#NUM!	0,48	-1,45	2,00	#NUM!	-1,28
50	0,00	0,00	#NUM!	#NUM!	0,48	0,62	2,00	#NUM!	-0,35

Table 6.1-1 Iteration sequence $z_{n+1} = z_n^2 - c$ for purely real c-values.

Nobody can carry out this kind of calculation with complex numbers in his head. For a few simple cases we will work out the boundary of the non–divergent region, before we try to obtain an overview of all possible c-values at once.

To begin with we will use purely real numbers c with no imaginary part. At each iteration step the current value is squared and c is subtracted. Thus the number only stays small if the square is not much larger than c. Furthermore, we will carry out the computation in a spreadsheet program.[2] In Table 6.1-1, we see the development for different real values of c.

In column 1 we find the variable n for $n = 0$ to 10. In order to get a glimpse of the later development, the values $n = 45$ to $n = 50$ are shown below. Next come the z_n-values, computed using the value of c listed in the first row.

Here is a brief commentary on the individual cases.

- Column 2, $c = 0$: z_n remains zero, no divergence.
- Column 3, $c = 1$: switching between $z = 0$ and $z = -1$, again no divergence.
- Column 4, $c = -1$: already after 5 steps the answers cannot be represented in the space available (#####) and after 15 steps they become larger than the biggest number with which EXCEL can work (#NUM!), namely 10^{200}. Certainly we have divergence here!

Now we must investigate the boundary between $c = -1$ and $c = 0$.

- $c = 0.5$ (column 5): surely not small enough, divergence.
- $c = -0.25$ (column 6): the first new case in which the c-values do not grow beyond reasonable bounds.
- The remaining case (columns 6–8) show that the upper boundary lies near $c = 2.0$.

Conclusion: the iteration sequence diverges if $c < -0.25$ or $c > 2.0$. In between we find (as in the Feigenbaum scenario in Chapter 2) simple convergence or periodic points, hence finite limiting values.

We have here carried out the iteration in great detail, in order to

- give you a feel for the influence of the c-values,
- show you how effectively a spreadsheet works,
- prepare you for the next step in the direction of complex numbers.

The investigation for purely imaginary parameters is not so easily carried out. Even when we square, we create from an imaginary number a negative real one, and upon subtracting we obtain a complex number! In Tables 6.1-2 and 6.1-3 we always list the real and imaginary parts of c and z next to each other. Otherwise both are constructed like Table 6.1-1.

---

[2]We use EXCEL, German version. As a result the decimal numbers have commas in place of the decimal point.

	Quadratic Iteration / c imaginary								
1	**2**	**3**	**4**	**5**	**6**	**7**	**8**	**9**	⬆
c →	0,00	0,50	0,00	-0,50	0,00	1,00	0,00	1,10	
n ↓	z-real	z-imag	z-real	z-imag	z-real	z-imag	z-real	z-imag	
0	0,00	0,00	0,00	0,00	0,00	0,00	0,00	0,00	
1	0,00	-0,50	0,00	0,50	0,00	-1,00	0,00	-1,10	
2	-0,25	-0,50	-0,25	0,50	-1,00	-1,00	-1,21	-1,10	
3	-0,19	-0,25	-0,19	0,25	0,00	1,00	0,25	1,56	
4	-0,03	-0,41	-0,03	0,41	-1,00	-1,00	-2,38	-0,31	
5	-0,16	-0,48	-0,16	0,48	0,00	1,00	5,55	0,35	
6	-0,20	-0,34	-0,20	0,34	-1,00	-1,00	30,66	2,83	
7	-0,08	-0,36	-0,08	0,36	0,00	1,00	931,80	172,70	
8	-0,13	-0,44	-0,13	0,44	-1,00	-1,00	*****	*****	
9	-0,18	-0,39	-0,18	0,39	0,00	1,00	*****	*****	
10	-0,12	-0,36	-0,12	0,36	-1,00	-1,00	*****	*****	⬇
45	-0,14	-0,39	-0,14	0,39	0,00	1,00	#NUM!	#NUM!	⬆
46	-0,14	-0,39	-0,14	0,39	-1,00	-1,00	#NUM!	#NUM!	
47	-0,14	-0,39	-0,14	0,39	0,00	1,00	#NUM!	#NUM!	
48	-0,14	-0,39	-0,14	0,39	-1,00	-1,00	#NUM!	#NUM!	
49	-0,14	-0,39	-0,14	0,39	0,00	1,00	#NUM!	#NUM!	
50	-0,14	-0,39	-0,14	0,39	-1,00	-1,00	#NUM!	#NUM!	⬇

Table 6.1-2 Iteration sequence for purely imaginary c-values.

- Columns 2 and 3 ($c = 0.5i$) show that a non-divergent sequence occurs.
- In columns 4 and 5 ($c = -0.5i$) we obtain the same numbers, except that the sign of the imaginary part is reversed.
- What we observe by comparing columns 2 and 3 with 4 and 5 can be expressed in a general rule: two conjugate complex[3] numbers c generate two sequences z which are always complex conjugates.
- In column 6 and 7 we see that $c = 1.0i$ provides an upper limit... .
- ... which is clear by comparing with columns 8 and 9 ($c = 1.1i$).

To see that the behaviour on the imaginary axis is not as simple as on the real axis, take a look at

- Columns 10 and 11: for $c = 0.9i$ the sequence diverges!

It is a total surprise that a small real part, for example

- $c = 0.06105 + 0.9i$ (columns 12 and 13),

again leads to an orderly (finite) sequence (at least within the first 50 iterations).

The last two examples, columns 14 and 15 ($c = 0.5+0.6i$) and 16 and 17 ($c = -0.3+0.5i$) should convince you that non-divergent sequences can be found when c is well removed from the axes. Indeed the number $-0.3+0.5i$ lies further to the left

---

[3]If you have difficulty with this concept, re-read the fundamental ideas of calculation with complex numbers in Chapter 4.

Quadratic Iteration / c imaginary								
1	**10**	**11**	**12**	**13**	**14**	**15**	**16**	**17**
c →	0,00	0,90	,06105	0,90	0,50	0,60	-0,30	0,50
n ↓	z-real	z-imag	z-real	z-imag	z-real	z-imag	z-real	z-imag
0	0,00	0,00	0,00	0,00	0,00	0,00	0,00	0,00
1	0,00	-0,90	-0,06	-0,90	-0,50	-0,60	0,30	-0,50
2	-0,81	-0,90	-0,87	-0,79	-0,61	0,00	0,14	-0,80
3	-0,15	0,56	0,07	0,47	-0,13	-0,60	-0,32	-0,72
4	-0,29	-1,07	-0,28	-0,84	-0,84	-0,45	-0,12	-0,04
5	-1,07	-0,28	-0,68	-0,43	0,01	0,15	0,31	-0,49
6	1,06	-0,30	0,22	-0,31	-0,52	-0,60	0,16	-0,81
7	1,03	-1,52	-0,11	-1,03	-0,58	0,02	-0,33	-0,75
8	-1,27	-4,03	-1,12	-0,68	-0,16	-0,63	-0,16	-0,01
9	-14,65	9,35	0,73	0,62	-0,87	-0,40	0,33	-0,50
10	127,30	*****	0,08	0,00	0,10	0,09	0,16	-0,82
45	#NUM!	#NUM!	0,25	0,36	0,03	0,15	0,32	-0,50
46	#NUM!	#NUM!	-0,13	-0,72	-0,52	-0,59	0,15	-0,82
47	#NUM!	#NUM!	-0,56	-0,72	-0,58	0,02	-0,35	-0,74
48	#NUM!	#NUM!	-0,26	-0,09	-0,17	-0,62	-0,13	0,02
49	#NUM!	#NUM!	0,00	-0,85	-0,85	-0,40	0,32	-0,50
50	#NUM!	#NUM!	-0,79	-0,90	0,07	0,08	0,15	-0,82

Table 6.1–3 Iteration sequence for purely imaginary and complex c-values.

than the boundary at $c = -0.25$, which have have discovered on the real axis.

We certainly will not obtain a complete overview using the tables – the problem is too complicated. We must resort to other means, and use graphical representation.

Every possible Julia set is characterised by a complex number. If we make the c-plane the basis of our drawing, each point in it corresponds to a Julia set. Since the point can be coloured either black or white, we can encode information there. As already stated, this is the information on the connectivty of the Julia set. If it is connected, a point is drawn at the corresponding screen position. If the appropriate Julia set is not connected, the screen remains blank at that point. In the following pictures, all points of the complex c-plane are drawn, for which that value of c belongs to the basin of the finite attractor. A program searches the plane point by point. In contrast to the Julia sets of the previous section, the initial value is fixed at $z_0 = 0$, while the complex number c changes.

The corresponding program is developed from Program Fragment 5.2–2. To start with, we have changed the name of the central functional procedure. It must throughout be called from the procedure Mapping. The above parameter will be interpreted as Creal and Cimaginary. We define x and y as new local variables, intialised to the value 0.

Program Fagment 6.1-1.

```
FUNCTION MandelbrotComputeAndTest (Creal, Cimaginary :
                real) : boolean;
   VAR
      iterationNo : integer;
      x, y, xSq, ySq, distanceSq : real;
      finished: boolean;
   PROCEDURE  StartVariableInitialisation;
   BEGIN
      finished := false;
      IterationNo := 0;
      x := 0.0; y := 0.0;
      xSq := sqr(x);  ySq := sqr(y);
      distanceSq := xSq + ySq;
   END; (* StartVariableInitialisation *)
   PROCEDURE compute;
   BEGIN
      IterationNo := IterationNo + 1;
      y := x*y;
      y := y+y-Cimaginary;
      x := xSq - ySq -Creal;
      xSq := sqr(x); ySq := sqr(y);
      distanceSq := xSq + ySq;
   END; (* compute *)

   PROCEDURE test;
   BEGIN
      finished := (distanceSq > 100.0);
   END; (* test *)

   PROCEDURE distinguish;
   BEGIN (* Does the point belong to the Mandelbrot set? *)
      MandelbrotComputeAndTest : =
         (IterationNo = MaximalIteration);
   END;  (* distinguish *)

   BEGIN (* MandelbrotComputeAndTest *)
      StartVariableInitialisation;
      REPEAT
         compute;
         test;
      UNTIL (IterationNo = MaximalIteration) OR finished;
```

```
    distinguish;
END;   (* MandelbrotComputeAndTest *)
```

The figure that this method draws in the complex plane is known as the *Mandelbrot set.*
To be precise, points belong to it when, after arbitrarily many iterations, only finite z-
values are produced. Since we do not have arbitrarily much time, we can only employ a
finite number of repetitions. We begin quite cautiously with 4 steps.

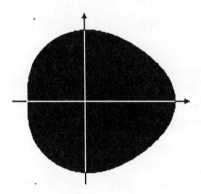

Figure 6.1-3 Mandelbrot set (4 repetitions).

As we see from the coordinate axes that are included, the resulting shape surrounds
the origin asymmetrically. In the calculation in Table 6.1-1 we have already noticed that
the basin stretches further in the positive direction.   After two further iterations the
egg–shaped basin begins to reveal its first contours.

Figure 6.1-4 and 5   Mandelbrot set (6, respectively 8 repetitions).

It gradually emerges that the edge of the figure is not everywhere convex, but that in
some places it undergoes constrictions.

The mirror symmetry about the real axis is obvious. It means that a point above the real axis has the same convergence behaviour as the corresponding point below. The complex numbers corresponding to these points are complex conjugates. In Table 6.1-2 we have already seen that such complex numbers have similar behaviour. They produce conjugate complex sequences.

Figures 6.1-6 and 7 Mandelbrot set (10 - 20 repetitions).

In Figure 6.1.6 some unusual points can be identified. The right-hand outermost point corresponds to $c = 2$. There the figure has already practically reduced to its final form: it exists only as a line along the real axis. However, magnified sections soon show that it possesses a complicated structure there.

The left upper branch of the same figure lies on the imaginary axis, where c has the value $c = i$. If we move from there directly downwards, we leave the basin of attraction, as we already know from Table 6.1-3 (columns 10 and 11).

Even if the drawing at first sight appears to show the opposite, the figure is connected. On the relatively large grid that determines the screen, we do not always encounter the extremely thin lines of which the figure is composed at many places. We already know this effect from the pictures of Julia sets.

All pictures can be combined and lead to 'contour lines'. In the next picture we not only draw the sets. In addition we draw the points, for which we can determine, after 4, 7, 10, 13, or 16 iterations, that they do not belong to it.

At this point we should describe the form of the Mandelbrot set. But, you may well ask, are there really similes for this uncommonly popular figure, which have not already been stated elsewhere?

In many people's opinion, it is reminiscent of a tortoise viewed from above, Randow (1986). Others see in it a remarkable cactus plant with buds, Clausberg (1986). For some mathematicians it is no more than a filled-in hypocycloid, to which a series of circles have been added, on which sit further circles, Durandi (1987). They are surely all correct, as also is the Bremen instructor who was reminded of a 'fat bureaucrat' (after rotation through 90°). Personal attitudes and interpretations certainly play a role when describing such an unusual picture. Two perceptions are involved, in our opinion. First, a familiarity with the forms that resemble natural shapes, in particular those seen upon

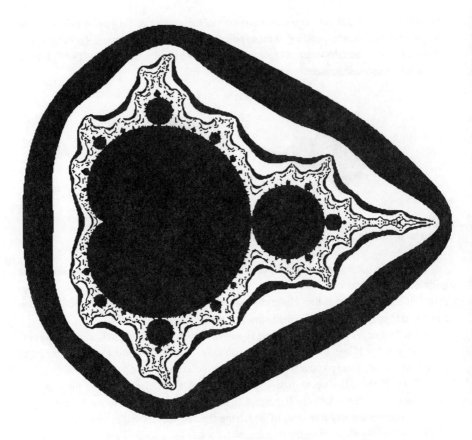

Figure 6.1-8 Mandelbrot set (100 repetitions, 'contour lines' up to 16).

magnification.   On the other hand, the strangeness of this type of repetition is very different from that of, e.g. a leaf and a tree. The structure on different scales, whereby more features of the shape emerge, contradicts our normal vision.   It makes us rethink our perceptions and eventually reach a new understanding.

In name-giving, and the fascination of form (and colour), we will follow the example of the Research Group at the University of Bremen, when in 1983 they were first able to produce the Mandelbrot set in their graphics laboratory.   The name 'Gingerbread Man'[4] arose spontaneously, and we find it so appropriate that we will also use it.

The figure itself is not much older than the name.   In the Spring of 1980 Benoît Mandelbrot first caught a glimpse of this graphic on a computer, hence its more formal

---

[4]*Translator's note*: 'Apfelmännchen' in the German, literally 'Little Apple Man'.   The 'translation' in the text, traditional among some sections of the English-speaking fractal community, captures the style of the German term. It is also a near-pun on 'Mandelbrot', which translates as 'almond bread'.

name.[5]

Never before has a product of esoteric mathematical research become a household word in such a short time, making numerous appearances on notice-boards, taking up vast amounts of leisure time,[6] and so rapidly becoming a 'superstar'.

In order to avoid misunderstandings in the subsequent description, let us define our terms carefully.[7]

- The complete figure, as in Figure 6.1-9, we will call the *Mandelbrot set* or *Gingerbread Man*. It is the basin of the 'finite attractor'.

- The approximations to the figure, as shown in Figure 6.1-8, are the *contour lines* or *equipotentialsurfaces*. They belong to the basin of the attractor ∞.

- In the Mandelbrot set we distinguish the *main body* and the *buds*, or baby Gingerbread Men.

- At some distance from the clearly connected central region we find *higher-order Mandelbrot sets* or *satellites*, connected to the main body by *filaments*. The only filament that can be clearly seen lies along the positive real axis. But there are also such connections to the apparently isolated points above and below the figure.

- We speak of *magnification* when not the whole figure, but only a section of it, is drawn. In a program we achieve this by selecting values for the variables Left, Right, Bottom, Top.

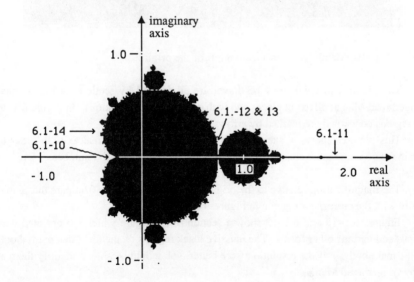

Figure 6.1-9 Mandelbrot set (60 repetitions).

[5]Compare the description in Peitgen and Richter (1986).

[6]If we can so describe the occupation with home and personal computers.

[7]They go back in particular to Peitgen and Richter (1986).

Figure 6.1-10 Mandelbrot set (section left of the origin).

The pictures that will now be drawn are all formed by sections of the original Gingerbread Man at different places and with different magnifications. In Figure 6.1-9, the regions concerned are marked by arrows.

This pictures shows a 'skewered Gingerbread Man' on the real axis. In order to obtain the contours more sharply, it is necessary to raise the iteration number. Here it is 100.

The magnification relative to the original is about 270-fold. Compare the central figure with the corresponding one in Figures 6.1-3 to 6.1-7.

Figures 6.1-12 and 6.1-13 show a section of the figure which is imprinted with spirals and various other forms. The massive black regions on the left side are offshoots of the main body. If the resolution were better, you would be able to identify them as baby Gingerbread Men.

Figure 6.1–11 A Mandelbrot set of the second order.

The final picture, Figure 6.1-14, has been turned 90°. In the original position the tiny Mandelbrot set has an orientation almost opposite to that of the original. It is attached to a branch with very strongly negative values for c_{real}. The magnification is about 500-fold.

The self-similarity observed here is also known in natural examples. Take a look at a parsley plant. For many 'iterations' you can see that two branches separate from each main stem. In contrast to this, the self-similarity of mathematical fractals has no limit. In a lecture, Professor Mandelbrot briefly showed a picture with a section of the Gingerbread Man magnified $6*10^{23}$ times (known to chemists as Avogadro's Number), which quite clearly still exhibited the standard shape.

Figure 6.1-12 A section betwen the main body and a bud.

Figure 6.1-13 A section directly below Figure 6.1-12.

Figure 6.1-14 A satellite fairly far left.

Computer Graphics Experiments and Exercises for §6.1

Exercise 6.1-1

Using a spreadsheet calculator or a Pascal program, construct an instrument to represent iterative sequences in tabular form.

Verify the results of Tables 6.1-1 to 6.1-3.

Find out what changes if instead of the formula

$$z_{n+1} = z_n^2 - c$$

you use

$$z_{n+1} = z_n^2 + c.$$

Exercise 6.1-2

To make the processes on the real axis, and their periodicity, more obvious, draw a Feigenbaum diagram for this case.   Draw the parameter c_{real} ($-0.25 \leq c_{\text{real}} \leq 2.0$) horizontally, and the z-value for 50 to 100 iterations vertically.

Exercise 6.1-3

Implement a Pascal program to draw the Mandelbrot set (Gingerbread Man).

Choose the limits for the region of the complex plane under investigation roughly as follows:

Left ≤ -1.0, Right ≥ 2.5 (Left $\leq c_{real} \leq$ Right)

Bottom ≤ -1.5, Top ≥ 1.5 (Bottom $\leq c_{imaginary} \leq$ Top).

For each screen pixel in this region start with the value

$$z_0 = x_0+i*y_0 = 0.$$

Draw all points, for which after 20 iterations the value of

$$|f(z)|^2 = x^2+y^2$$

does not exceed the bound 100.

It should take your computer about an hour to do this.

Exercise 6.1-4

When you have plenty of time (e.g. overnight), repeat the last exercise with 50, 100, or 200 steps. The more iteration steps you choose, the more accurate the contours of the figure will be.

Draw 'contour lines' too, to make the approximations to the Mandelbrot set more accurate.

Exercise 6.1-5

Investigate sections of the Mandelbrot set, for which you choose the values for Left, Right, Bottom, Top yourself.

Take care that the real section (Right - Left) and the imaginary section (Top - Bottom) always stay in the same proportion as the horizontal (Xscreen) and the vertical (Yscreen) dimensions of your screen or graphics window. Otherwise the pictures will appear distorted, and this strangeness can detract from their pleasing appearance. If you do not have a 1:1 mapping of the screen onto the printer output, you must also take care of this. For large magnifications you must also increase the iteration number.

In the pictures in this section you can already see that it is the boundary of the Mandelbrot set that is graphically the most interesting.

Worthwhile objects of investigation are:

- The filament along the real axis, e.g. the neighbourhood of the point $c = 1.75$,
- Tiny 'buds' that sprout from other larger ones,
- The regions 'around' the buds,
- The valleys between the buds,
- The apparently isolated satellites, which appear some distance from the main body, and the filaments that lead to them.

By now you will probably have found your own favourite regions within the Mandelbrot set. The boundary is, just like any fractal, infinitely long, so that anyone can explore new territory and generate pictures absolutely unseen before.

Exercise 6.1-6

The assumption that all iterations begin with

$$z_0 = x_0 + i*y_0 = 0$$

is based on a technical requirement. For our graphical experiments it is not a necessity. Therefore, try constructing sequences beginning with a number other than zero, with $x_0 \neq 0$ and/or $y_0 \neq 0$.

Exercise 6.1-7

The iteration sequence studied in this section is of course not the only one possible. Perhaps you would like to reconstruct the historical situation in the Spring of 1980, which B. B. Mandelbrot described in his article. Then you should try using the sequence

$$z_{n+1} = c*(1+z_n^2)^2/(z_n^2*(z_n^2-1))$$

or

$$z_{n+1} = c*z_n*(1-z_n).$$

The boundaries and the remaining parameters you must find out for yourself by experimenting. Enjoy yourself: it will be worth it!

6.2 Tomogram of the Gingerbread Man

Julia sets and Mandelbrot sets, such a variety of forms, such fantastic patterns! And all that happens if you just iterate a simple nonlinear equation with a complex parameter.

If we resolve the possible parameters into their components, we find that we can influence the iteration mathematically in four places. The first two values available to us are the real and the imaginary component of the initial value z_0; the other two are the components of c.

Because these four quantities can be varied independently of one another, we obtain a fourfold range of new computational foundations and thus a fourfold system of new pictures, when we combine the quantities in different ways. The true structure of the attractor of the iteration formula

$$z_{n+1} = z_n^2 - c$$

is four-dimensional!

Most people experience severe difficulties thinking about three–dimensional situations. Everything that transcends the two dimensions of a sheet of paper discloses itself only with great difficulty, and only then when one has much experience with the subject under investigation.   There is no human experience of four independent directions, and four mutually orthogonal coordinate axes cannot be represented artistically or technically.   If people wish, despite this, to get a glimpse of higher–dimensional secrets, there is only one possibility: to make models that reduce the number of dimensions. Every architect or draughtsman reduces the number of dimensions of his real objects from three to two, and can thus put them on paper. Each photo, each picture

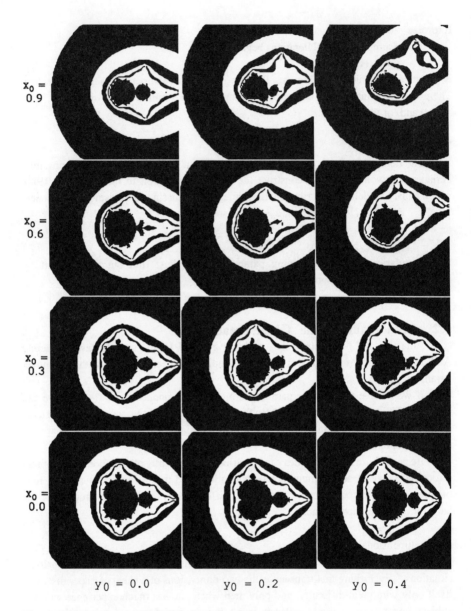

Figure 6.2-1   Quasi–Mandelbrot sets for different initial values.

basically does the same.

A particularly pretty example appears on the cover of Douglas R. Hofstadter's book *Gödel, Escher, Bach*. A shape, presumably cut from wood, has a shadow which from

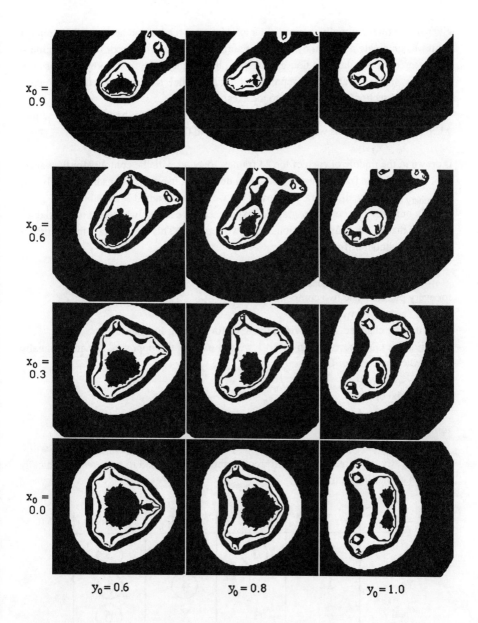

$x_0 = 0.9$

$x_0 = 0.6$

$x_0 = 0.3$

$x_0 = 0.0$

$y_0 = 0.6$ $y_0 = 0.8$ $y_0 = 1.0$

Figure 6.2-2 Quasi-Mandelbrot sets for different initial values.

different directions looks like the letters G, E, or B. Thus we see that such a reduction of dimension can colloquially be described as a *silhouette*, a *section*, or a *tomogram*. Architects and mathematicians alike build upon a three-dimensional coordinate system given by the three mutually perpendicular axes length, breadth, and height. In its simplest form we can create a section by fixing the numerical value in one coordinate direction – for example, in the plan of a house, looking at the height of the first storey. In this case the space that is graphically represented runs parallel to the two remaining axes. Complicated sections run obliquely to the axes.

The pictures on the previous two pages survey the form of the basin of attraction, when the iteration begins with the value shown round the edges. That is,

$$z_0 = x_0 + i y_0$$

is drawn in the middle of each of the 24 frames, together with the contours for the values 3, 5, and 7. In the frames, just as in the Mandelbrot set itself, c_{real} runs horizontally and $c_{imaginary}$ vertically. In the frame at lower left we see the standard Mandelbrot set.

In our previous computer graphics experiments we have always kept two variables fixed, so that of the four dimensions only two remain. This lets us draw the results on the screen without difficulty.

In §5.2 we fixed

$$c = c_{real} + i * c_{imaginary}$$

for every picture. The two components of

$$z_0 = x_0 + i * y_0$$

could then be changed, and provided the basis for drawing Julia sets.

For the Gingerbread man in §6.1 we did exactly the opposite, a mathematically 'perpendicular' choice. There

$$z_0 = x_0 + i * y_0$$

remained fixed, while

$$c = c_{real} + i * c_{imaginary}$$

formed the basis of the computation and the drawing.

Building on the four independent quantities x_0, y_0, c_{real}, $c_{imaginary}$ we will systematically investigate which different methods can be used to represent graphically the basin of the finite attractor.

	x_0	y_0	c_{real}	$c_{imaginary}$
x_0	X	(1)	(2)	(3)
y_0	1	X	(4)	(5)
c_{real}	2	4	X	(6)
$c_{imaginary}$	3	5	6	X

Table 6.2-1   Possibilities for representation.

In Table 6.2-1 we show the $4 \times 4 = 16$ possibilities, which can be expressed using two of these four parameters. The two quantities on the upper edge and down the side are kept constant in the graphics. The remaining two are still variable, and form the basis of the drawing. Because of the 'four-dimensional existence' of the basin of attraction, we select sections that are parallel to the four axes.

The four possibilities along the diagonal of Table 6.2-1 do not give a sensible graphical representation, because the same quantity occurs on the two axes. Corresponding to each case above the diagonal there is one below, in which the axes are interchanged. For the basic investigation that we carry out here, this makes no difference. Thus there remain 6 distinct types, with which to draw the basin of the iteration sequence.

By case 1 we refer to that in which x_0 and y_0 are kept fixed. Then c_{real} and $c_{imaginary}$ are the two coordinates in the plane that underlies each drawing. A special case of this, which we call case 1a, is when $x_0 = y_0 = 0$, and this leads to pictures of the Mandelbrot set or Gingerbread Man. Case 1b, for which $x_0 \neq 0$ and/or $y_0 \neq 0$, has already been worked out in Exercise 6.1-6. A general survey of the forms of the basins may be found in Figs. 6.2-1 and 6.2-2. The intial values x_0 and y_0 are there chosen from the range 0 to 1. Lacking any better name we call them *quasi-Mandelbrot sets*.

We recommend you to find out what happens when one or both components of the initial value are negative. Do you succeed in confirming our previous perception that the pictures of the basins becomes smaller and more disconnected, the further we go away from the starting value

$$z_0 = x_0 + i*y_0 = 0?$$

And that we obtain symmetric pictures only when one of the two components x_0 or y_0 has the value zero?

Another case, already used in §5.2 as a method for constructing Julia sets, is number 6 in the Table.

If you have already wondered why the Julia sets and the Gingerbread Man have so little in common, perhaps a small mathematical hint will help. In a three-dimensional space there are three possible ways in which two planes can be related. They are either equal, or parallel, or they cut in a line. In four-dimensional space there is an additional possibility: they 'cut' each other in a point. For Julia sets this is the origin. For the Gingerbread Man this is the parameter c, which is different for each Julia set.

The two real quantities x_0 and c_{real} are kept fixed throughout. The basis for the drawing is then the two imaginary parts y_0 and $c_{imaginary}$. The pictures are symmetric about the origin, which is in the middle of each frame.

The central basin is rather small near $c_{real} = 0.8$ Recall that at $c = 0.75$ we find the first constriction in the Gingerbread Man, where its dimension along the imaginary axis goes to zero. And this axis is involved in the drawing.

The central basin is fairly shapeless for small values of x_0 and $c_{imaginary}$. Perhaps this is just a matter of the depth of iteration, which for all the pictures here is given by maximalIteration = 100.

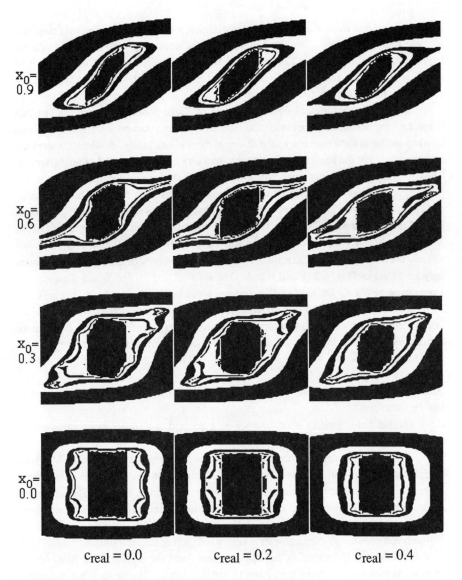

Figure 6.2-3 Diagram for case 2.

Interesting forms first arise for values such that the sets are rapidly disappearing. There it also looks as though the basin is no longer connected.

The pictures on the surrounding pages are symmetric about the origin. They all lie in the x_0-$c_{imaginary}$ plane, with $c_{imaginary}$ being drawn vertically. We also find this type of symmetry in Julia sets.

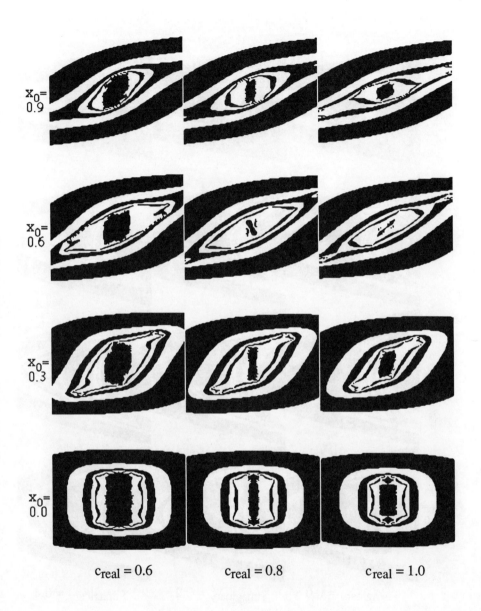

Figure 6.2-4 Diagram for case 2.

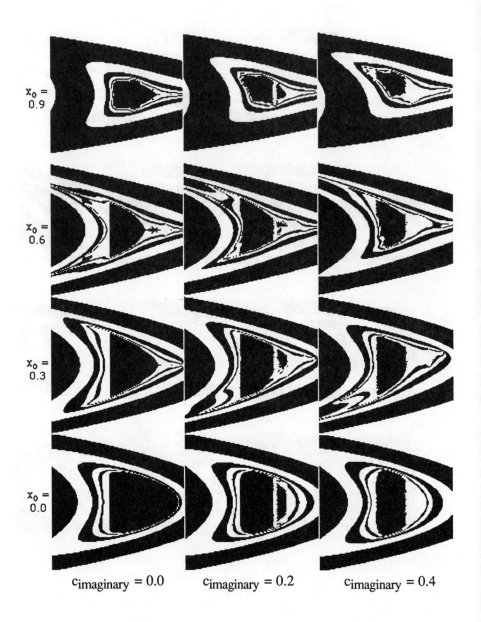

Figure 6.2-5 Diagram for case 3.

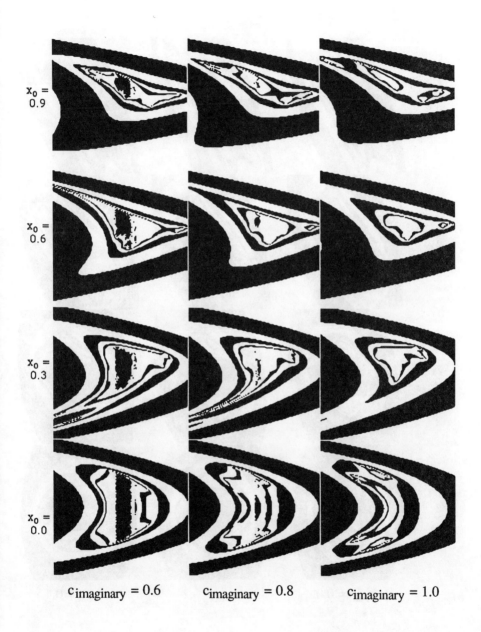

Figure 6.2–6 Diagram for case 3.

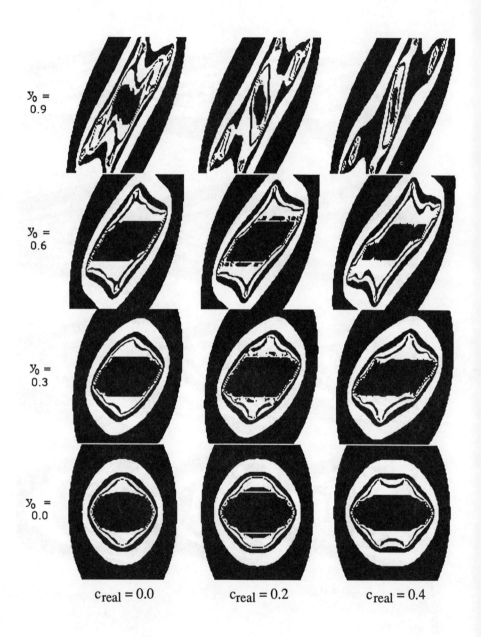

Figure 6.2–7 Diagram for case 4.

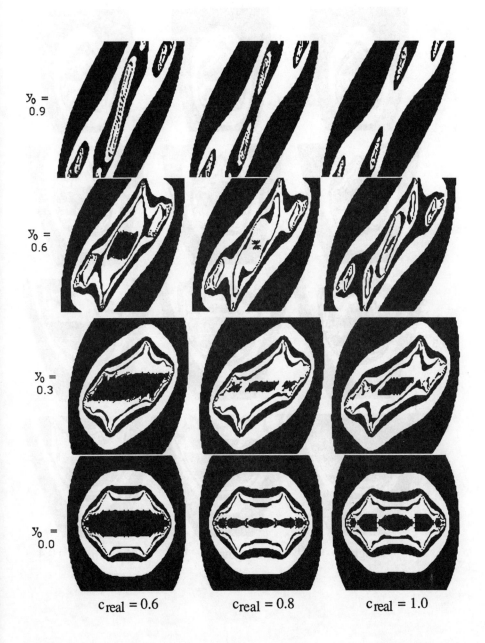

Figure 6.2-8 Diagram for case 4.

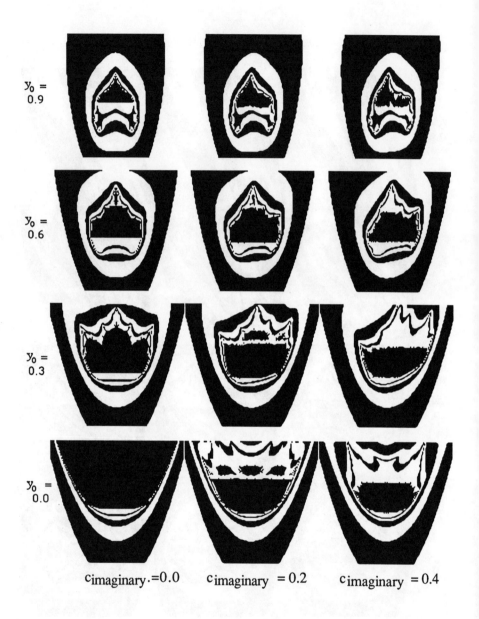

Figure 6.2-9 Diagram for case 5.

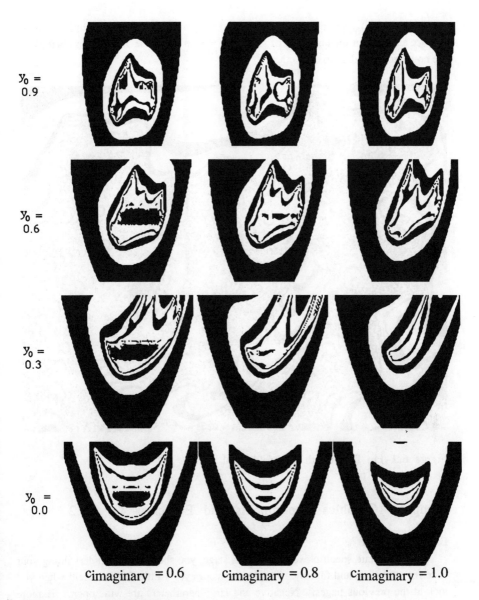

$y_0 = 0.9$

$y_0 = 0.6$

$y_0 = 0.3$

$y_0 = 0.0$

$c_{imaginary} = 0.6$ $c_{imaginary} = 0.8$ $c_{imaginary} = 1.0$

Figure 6.2-10   Diagram for case 5.

It is clear that the central basin is no longer connected.   If you produce magnifications of the details, you must take into account the extreme irregularity (see Exercise 6.2-1). If we keep the two imaginary quantities y_0 and $c_{imaginary}$ constant, and in each picture change x_0 and c_{real}, we get the most remarkable and wildest shapes.

Provided $c_{imaginary} > 0$, the sets break up into many individual regions. In Fig. 6.2-11 you see a section of the boundary of such a set, drawn as in case 5. Its individual pieces are no longer joined up to each other.

Figure 6.2-11   Detail with sevenfold spiral.

Computer Graphics Experiments and Exercises for §6.2

Exercise   6.2-1

To formulate specific exercises at this stage, you should avoid undervaluing your taste for adventure and fantasy. Simply cast your eyes a little further afield when you look at the previous pages. Negative and large parameters are wide open. Explore sections of the pictures shown. In some cases you must then employ different scales on the two axes, or else the pictures will be distorted.

We show you an attractive example above in Fig. 6.2-11. The two values $y_0 = 0.1$ and $c_{imaginary} = 0.4$ are fixed. From Left to Right the real starting values changes in the range $0.62 \leq x_0 \leq 0.64$. From Bottom to Top c_{real} varies: $0.74 \leq c_{real} \leq 0.8$. In the diagonal direction the original square figure is stretched by a factor of 3.

Exercise 6.2-2

Experiment with 'slanting sections'. From left to right change both c_{real} and $c_{imaginary}$ using a formula of the type

$$c_{real} = 0.5 * c_{imaginary}$$

or similar. Or nonlinear expressions. From bottom to top x_0, say, varies, while y_0 stays fixed.

6.3 Fig-tree and Gingerbread Man

The definitive investigations that we have carried out for the Gingerbread Man in §6.1 have reminded us that in different regions of the complex plane it is possible to find different periodicities in the number sequences. It is therefore not so easy to distinguish between periodic behaviour and chaos – for instance, when the period is large. The highest number in respect of which we will consider periodicity is therefore 128. For purely computational procedures a whole series of problems arise. First, we must wait several hundred iterations before the computation 'settles down'. By this we mean it reaches a stage at which order and chaos can be distinguished. The next difficulty is that of comparison. The internal computer code for representing real numbers in Pascal is not completely unequivocal,[8] so that equality can only be tested using a trick. Thus we can investigate only whether the numbers differ by less than an assigned bound (e.g. 10^{-6}). In Pascal we can formulate a functional procedure like that in Program Fragment 6.3-1:

Program Fragment 6.3-1

```
FUNCTION equal (no1, no2: real) : boolean;
BEGIN
    equal := (ABS(no1-no2)) < 1.0E-6);
END;
```

For complex numbers, we must naturally check that both the real and the imaginary parts are equal:

Program Fragment 6.3-2

```
FUNCTION equal (z1Real, z1Imag, z2Real, z2Imag : real) :
            boolean;
BEGIN
    equal := (ABS(z1Real-z2Real)+ABS(z1Imag-z2Imag)
            < 1.0E-6;
END;
```

---

[8] The same sort of thing happens with the decimal numbers 0.1 and 0.0999..., which are equal, but written differently.

For example, if we have established that the 73rd number in a sequence is equal to the 97th, then we can conclude that a period of 24 then exists. This condition can be used to colour the uniformly black region inside the Gingerbread man.

We offer this as an example, and collect the results together:

- The constrictions in the Mandelbrot set (or, more poetically, the necks of the Gingerbread Man) divide regions of different periodicity from each other.
- The main body has period 1, that is, each number sequence $z_{n+1} = z_n^2 - c$ beginning with $z_0 = 0$, for c chosen within this region, tends towards a fixed complex limit. If c is purely real, so is the limit.
- The first circular region, which adjoins it to the right, leads to sequences of period 2.
- The further 'buds' along the real axis exhibit the periods 4, 8, 16, ...
- We find period 3 in the two next largest adjoining buds near the imaginary axis, near $c = 1.75$.
- For each further natural number we can find closed regions of the Mandelbrot set in which that periodicity holds.
- Regions that adjoin each other differ in periodicity by an integer factor. Adjacent to a region with periodicity 3 we find regions with the periods 6, 9, 12, 15, etc. The factor 2 holds for the largest bud, the factor 3 for the next largest, and so on.
- At the limits of the above regions the convergence of the number sequence becomes very poor, so that we need far more than 100 iterations to decide the question of periodicity.

We have already had a lot to do with this condition in Chapter 2, in the study of the Feigenbaum phenomenon. First, as a bridge–building exercise, we show that the new graphics display a similar state of affairs. Thus even the formula that underlies the Mandelbrot set can be drawn in the form of a Feigenbaum diagram.

The parameter to be varied here is, in the first instance, c_{real}. To begin with, the imaginary part $c_{imaginary}$ will be held at a constant value of 0. The program that we used in §6.1 to draw the Mandelbrot set changes a little.

To make as few alterations to the program as possible, we relinquish the use of the global variables Visible and Invisible. Their role will be played by boundary and maximalIteration, respectively.

Program Fragment 6.3-3

```
PROCEDURE Mapping;
   VAR
       Xrange : integer;
       deltaXPerPixel : real;
       dummy : boolean;
```

```
FUNCTION  ComputeAndTest (Creal, Cimaginary : real) :
                boolean;
    VAR
       IterationNo : integer;
       x, y, xSq, ySq, distanceSq : real;
       finished : boolean;

    PROCEDURE StartVariableInitialisation;
    BEGIN
       x := 0.0; y := 0.0;
       finished := false;
       iterationNo := 0;
       xSq := sqr(x); ySq := sqr(y);
       distanceSq := xSq + ySq;
    END; (* StartVariableInitialisation *)

    PROCEDURE ComputeAndDraw;
    BEGIN
       IterationNo := IterationNo + 1;
       y := x*y; y := y+y-Cimaginary;
       x := xSq - ySq - Creal;
       xSq := sqr(x); ySq := sqr(y);
       distanceSq := xSq + ySq;
       IF (IterationNo > Bound) THEN
                SetUniversalPoint  (Creal,x);
    END; (* ComputeAndDraw *)

    PROCEDURE test;
    BEGIN
       finished := (distanceSq > 100.0);
    END; (* test *)

  BEGIN (* ComputeAndTest *)
     StartVariableInitialisation;
     REPEAT
        computeAndDraw; test;
     UNTIL (IterationNo = MaximalIteration) OR finished;
  END (* ComputeAndTest *)

BEGIN
   deltaxPerPixel := (Right - Left)/Xscreen;
```

```
. x := Left;
  FOR xRange := 0 TO Xscreen DO
  BEGIN
     dummy := ComputeAndTest (x, 0.0);
     x := x + deltaxPerPixel;
  END;
END; (* Mapping *)
```

As you see, we now need a loop which increments the running variable xRange. As a result, variables for the other loops become superfluous. A new introduction is the variable dummy. By including it we can call the functional procedure ComputeAndTest as before. The drawing should be carried out during the iteration, and would be called from Mapping and built into ComputeAndTest.

As global parameters we use

```
Left := -0.25; Right := 2.0; Bottom := -1.5; Top := 1.5;
MaximalIteration := 300; Bound := 200;
```

and away we go!

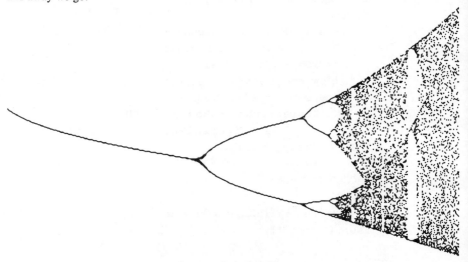

Figure 6.3-1 Feigenbaum diagram from the Mandelbrot set.

The result in Figure 6.3-1 appears very familiar, when we recall Chapter 2. It shows that the Feigenbaum scenario with bifurcations, chaos, and periodic windows is present in all respects along the real axis of the Mandelbrot set.

The next figure, 6.3-2, illustrates this by drawing the two diagrams one above the other: the periods 1, 2, 4, 8, ... etc. can be identified in the (halved) Mandelbrot set just as well as in the Feigenbaum diagram. And the 'satellite' Gingerbread Man corresponds in the Feigenbaum diagram to a periodic window with period 3! The diagram is only defined where it lies within the basin of the finite attractor, $-0.25 \le c_{real} \le 2.0$. For other values of c_{real} all sequences tend to $-\infty$.

Figure 6.3-2 Direct comparison: Gingerbread Man and Fig-tree.

Now the real axis is a simple path, but not the only one along which a parameter can be changed. Another interesting path is shown in Figure 6.3-3.

$\leftarrow$ In the diagrams, c varies along this line

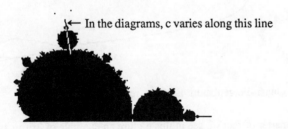

Figure 6.3-3 A parameter path in the Mandelbrot set.

This path is interesting because it leads directly from a region of period 1 to a bud of period 3, then on to 6 and 12.

This straight line is specified by two points, for example those at which the buds᾽ touch. Using a program to investigate the Mandelbrot set we have discovered the coordinates of the points P_1 (0.1255, 0.6503) between the main body and the first bud, and P_2 (0.1098, 0.882) where the next bud touches. The equation of the line can then be specified in two–point form. Using the parameter xRange, which runs from 0 to 400, we can travel along the desired section.

Mapping then takes the following form:

Program Fragment 6.3-4

```
(Working part of the procedure Mapping)
BEGIN (* Mapping *)
   FOR xRange := 0 TO xScreen DO
      dummy := ComputeAndTest
         (0.1288 - xRange*6.767E-5,          {Creal}
          0.6 + xRange*1.0E-3);              {Cimaginary}
   END ; (* Mapping *)
```

Now we must clarify what should actually be drawn, since ultimately both the real and imaginary parts of *z* are to be studied. We try both in turn. To draw the real part *x*, the appropriate part of ComputeAndTest runs like this:

Program Fragment 6.3-5 (Drawing commands in ComputeAndTest)

```
IF (IterationNo > Bound) THEN
      SetUniversalPoint (Cimaginary, x);
```

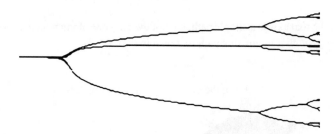

Figure 6.3-4 Quasi–Feigenbaum diagram, real part.

And as a matter of fact we see in this picture an example of 'trifurcation', when the period changes from 1 to 3. It looks rather as if first two paths separate, and then a further one branches off these. But that is just an artefact of our viewpoint.

In the next figure we look at the imaginary part *y*. For Figure 6.3-5 the relevant commands run like this:

Program Fragment 6.3-6 (Drawing commands in ComputeAndTest)

```
IF (IterationNo > Bound) THEN
      SetUniversalPoint (Cimaginary, y);
```

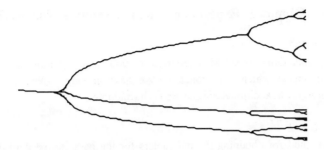

Figure 6.3-5 Quasi-Feigenbaum diagram, imaginary part.

We obtain a fairly complete picture using a pseudo-three-dimensional representation, as follows:

Program Fragment 6.3-7 (Drawing commands in `ComputeAndTest`)

```
IF (IterationNo > Bound) THEN
    SetUniversalPoint (Cimaginary - 0.5*x, y+0.866*x);
```

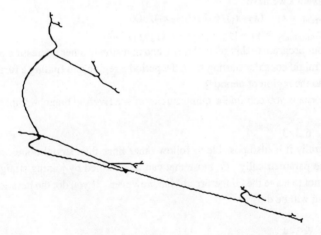

Figure 6.3-6 Pseudo-three-dimensional representation of trifurcation (oblique view from the front).

Here we get a complete view of the Quasi-Feigenbaum diagram. Two of the three main branches are shortened by perspective. The 24 tiny twigs appear at greater heights to come nearer the observer, whereas the first few points of the figure (period 1) are towards the rear. We can clearly see that the threefold branching occurs at a single point.

Computer Graphics Experiments and Exercises for §6.3

Exercise 6.3-1

Using your Gingerbread Man program, or your own variation on it, find other boundary points at which buds touch. We have already looked at the thinnest constriction and done additional computation at its midpoint.

Exercise 6.3-2

The method for obtaining the parameters for the path can be described here in general terms.[9]

We have found the two points P_1 (x_1,y_1) and P_2 (x_2,y_2). Thus, a 400-pixel line between them can be seen schematically as follows:

```
         P1                                      P2
--------+------------------------------------+----------
 0     50                                   350     400
```

Let the variable t run along this line (in the program t becomes xRange). Then for the two components we have

$$c_{real} = x_1 - (x_2-x_1)/6 + t*(x_2-x_1)/300,$$
$$c_{imaginary} = y_1 - (y_2-y_1)/6 + t*(y_2-y_1)/300.$$

Change your program to this general form, and investigate other interesting paths.

You might consider starting from the period 3 region and finding a further sequence leading into the region of period 9.

Or perhaps you can find a 'quintufurcation' - a fivefold branch-point?

Exercise 6.3-3

Naturally it is also possible to follow other lines than straight ones, which can be represented parametrically. Or those that can be obtained by joining straight segments. Take care not to leave the Gingerbread Man, however. If you do, the iteration will cease, and nothing will be drawn.

Exercise 6.3-4

We have obtained the pseudo-3D effect using the following trick.

In principle we have drawn the $c_{imaginary}$ diagram. But we have added to or subtracted from each of the two components a multiple of the x-value. The numbers 0.5 and 0.866 come from the values of the sine and cosine of 30°.

Experiment using other multiples.

---

[9]In program Fragment 6.3-4 we have proceeded somewhat differently.

6.4   Metamorphoses

With the Gingerbread Man and Julia sets constructed from the quadratic feedback mapping

$$z_{n+1} = z_n^2 - c$$

we have reached some definite conclusions.   For investigations that go beyond these we can only provide a few hints.   The questions that arise are so numerous that even for us they are for the most part unsolved and open.   Treat them as problems or as further exercises.

First of all, we should point out that the quadratic mapping is not the only possible form for feedback.   To be sure, it is the simplest that leads to 'nontrivial' results.   In this book we have tried to avoid equations of higher degree, for which the computing time increases steeply, and we have only done so when it illustrates worthwhile principles.

In Peitgen and Richter (1986) p. 106, the authors describe the investigation of rational (that is, fractional) functions, which occur in physical models of magnetism.   As above, there is a complex variable z which is iterated, and a constant c which in general is complex.   The appropriate equations are:

$$z_{n+1} = \left(\frac{z_n^2 + c - 1}{2z_n + c - 2} \right)^2 \qquad \text{Model 1,}$$

$$z_{n+1} = \left(\frac{z_n^3 + 3(c-1)z_n + (c-1)(c-2)}{3z_n^2 + 3(c-2)z_n + c^2 - 3c + 3} \right)^2 \qquad \text{Model 2.}$$

Again there are two methods of graphical representation.   We draw in either the z-plane or the c-plane, so that in the first case we choose a fixed c-value, and in the second case we begin with $z_0 = 0$.   You should decide for yourself the c-values, which can also be real, the boundaries of the drawings, and the type of colouring used.   Of course there is nothing wrong in experimenting with other equations or modifications.

We encounter rational functions if we pursue an idea from Chapter 4.   There we applied Newton's method to solve a simple equation of the third degree.   It can be shown (see Curry, Garnett and Sullivan, 1983) that we can investigate similar cubic equations with the formula

$$f(z) = z^3 + (c-1)*z - c.$$

Here c is a complex number.

We begin the calculations once more with

$$z_0 = 0,$$

insert different c-values, and apply Newton's method:

$$z_{n+1} = z_n - \frac{f(z_n)}{f'(z_n)} = z_n - \frac{z_n^3 + (c-1)z_n + c}{3z_n^2 + c - 1} .$$

We find three different types of behaviour for this equation, depending on c. In many cases, in particular if we calculate with c-values of large modulus, the sequence converges to the real solution of the equation, namely

$z = 1$.

In other cases the sequence converges to another root. In a few cases the Newton method breaks down completely. Then we get cyclic sequences, that is, after a number of steps the values repeat:

$z_{n+h} = z_n$,

where h is the length of the cycle.

We can draw all of the points in the complex c-plane corresponding to the first case, using a Pascal program.

The calculation requires so many individual steps that it can no longer, as in the previous examples, be programmed 'at a walking pace'. For that reason we provide here a small procedure for calculating with complex numbers. The complex numbers are throughout represented as two 'real' numbers.

The components deal with addition, subtraction, multiplication, division, squaring, and powers. All procedures are constructed in the same way. They have one or two complex input variables (in1r stands for 'input 1 real part' etc.) and one output variable as VAR parameter.

In respect of division and powers we must take care of awkward cases. We do not, for example, consider it sensible to stop the program if we inadvertently divide by the number zero. Thus we have defined a result for this value too.

In your program these procedures can be defined globally, or locally within ComputeAndTest.

Program Fragment 6.4-1

```
PROCEDURE compAdd (in1r, in1i, in2r, in2i: real; VAR outr,
                   outi: real);
BEGIN
   outr := in1r + in2r;
   outi := in1i + in2i;
END;   (* compAdd *)


PROCEDURE compSub (in1r, in1i, in2r, in2i: real; VAR outr,
                   outi: real);
BEGIN
   outr := in1r - in2r;
   outi := in1i - in2i;
END;   (* compSub *)


PROCEDURE compMul (in1r, in1i, in2r, in2i: real; VAR outr,
```

```
                         outi: real);
BEGIN
   outr := in1r * in2r - in1i * in2i;
   outi := in1r * in2i + in1i * in2r;
END;   (* compMul *)


PROCEDURE compDiv (in1r, in1i, in2r, in2i: real; VAR outr,
                     outi: real);
   VAR numr, numi, den: real;
BEGIN
   compMul (in1r, in1i, in2r, -in2i, numr, numi);
   den := in2r * in2r + in2i * in2i;
   IF den := 0.0 THEN
      BEGIN
         outr := 0.0; outi := 0.0;   (* emergency solution *)
      END
   ELSE
      BEGIN
         outr := numr/den;
         outi := numi/den;
      END;
END;   (* compDiv *)


PROCEDURE compSq (in1r, in1i : real; VAR outr, outi: real);
BEGIN
   outr := in1r * in1r - in1i * in1i;
   outi := in1r * in1i * 2.0;
END;   (* compSq *)


PROCEDURE compPow (in1r, in1i, power: real; VAR outr, outi:
                     real);
   CONST
      halfpi := 1.570796327;
   VAR
      alpha, r : real;
BEGIN
   r := sqrt (in1r*in1r + in1i * in1i);
   IF  r > 0.0 then r := exp (power * ln(r));
   IF ABS(in1r) < 1.0E-9 THEN
      BEGIN
         IF in1i > 0.0 THEN alpha := halfpi;
```

```
                                   ELSE alpha := halfpi + Pi;
          END ELSE BEGIN
             IF inlr > 0.0 THEN alpha := arctan (inli/inlr)
                             ELSE alpha := arctan (inli/inlr) + Pi;
          END;
       IF alpha < 0.0 THEN alpha := alpha + 2.0*Pi;
       alpha := alpha * power;
       outr := r * cos(alpha);
       outi := r * sin(alpha);
    END;   (* compPow *)
```

Having equipped ourselves with this utility we can now carry out an investigation of the complex plane. Replace the functional procedure `MandelbrotComputeAndTest` in your Gingerbread Man program by one based upon the above procedures. But do not be surprised if the computing time becomes a bit longer.

Program Fragment 6.4-2 (Curry–Garnett–Sullivan Method)

```
   FUNCTION ComputeAndTest (Creal, Cimaginary : real) :
                     boolean;
   VAR
       IterationNo : integer;
       x, y, distanceSq, intr, inti, denr, deni : real;
       (* new variables to store the denominator *)
       (* and intermediate results *)
       finished : boolean;

   PROCEDURE StartVariableInitialisation;
   BEGIN
       finished := false;
       IterationNo := 0;
       x := 0.0;
       y := 0.0;
   END; (* StartVariableInitialisation *)

   PROCEDURE compute;
   BEGIN
       IterationNo := IterationNo + 1;
       compSq (x, y, intr, inti);
       compAdd (3.0*intr, 3.0*inti, Creal-1.0, Cimaginary,
                   denr, deni);
       compAdd (intr, inti, cReal -1.0, Cimaginary, intr,
                   inti);
```

```
      compMul (intr, inti, x, y, intr, inti);
      compSub (intr, inti, Creal, Cimaginary, intr, inti);
      compDiv (intr, inti, denr, deni, intr, inti);
      compSub (x, y, intr, inti, x, y);
      distanceSq := (x-1.0) * (x-1.0) + y * y;
   END;   (* compute *)


   PROCEDURE test;
   BEGIN
      finished := (distanceSq < 1.0E-3);
   END (* FurtherTest *)


   PROCEDURE distinguish;
   BEGIN (* does the point belong to the set? *)
      ComputeAndTest := iterationNo < maximalIteration;
   END;   (* distinguish *)

BEGIN (* ComputeAndTest *)
   StartVariableInitialisation;
   REPEAT
      compute;
      test;
```

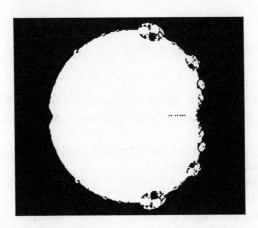

Figure 6.4-1 Basin of the attractor $z = 1$.

Figure 6.4-2 Section from Figure 6.4-1 (with a surprise!).

```
UNTIL (IterationNo = MaximalIteration) OR finished;
distinguish;
END;  (* ComputeAndTest *)
```

As you see, the computing expense for each step has grown considerably. As a result you should not choose too large a number of iterations.

Figure 6.4-1 has a very clear structure and displays some interesting regions which are worth magnifying. For example, you should investigate the area around

$c = 1,$

$c = 0,$

$c = -2.$

The elliptical shape in the neighbourhood of

$c = 1.75i$

is enlarged in Figure 6.4-2.

- Black areas correspond to regions in which z_n converges to $z = 1$.
- Most white regions mean that case 2 holds there. The sequence converges, but not to $z = 1$.
- A white region at the right-hand end of the figure is an exception. This is a region where the sequences become cyclic.

Check this for c-values near

$c = 0.31 + i*1.64.$

You can already discern the result in Figure 6.4-2.   In fact what you get is a close variant of the Gingerbread Man from §6.1.

This resemblance to the Gingerbread Man is of course no accident. What we named the 'finite attractor' at the start of Chapter 5 is mostly a cyclic attractor, as in this computation.

If the change to third powers already leads to such surprising results, what will we find for fourth or even higher powers? To get at least a few hints, we generalise the simple iteration equation

$$z_{n+1} = z_n{}^2 - c.$$

Instead of the second power we use the pth:

$$z_{n+1} = z_n{}^p - c.$$

We carry out the power computation using the procedure compPow (Program Fragment 6.4-1), which involves the global variable p. The changes to the Gingerbread man program are limited to the procedure Compute.

Program Fragment 6.4-3

```
PROCEDURE compute;
    VAR
        tempr, tempi: real;
BEGIN
    IterationNo := IterationNo + 1;
    compPow (x, y, p, tempr, tempi);
    x : = tempr - Creal;
    y := tempi - Cimaginary;
    xSq := sqr(x);
    ySq := sqr(y);
    distanceSq := xSq + ySq;
END;   (* compute *)
```

A brief assessment of the results shows that for the power $p = 1$ the finite attractor is confined to the origin. Every other c-value moves further away, that is, to the attractor ∞.

For very high values of p we can guess that the modulus of c plays hardly any role compared with the high value of z^p, so that the basin of the finite attractor is fairly close to the unit circle. Inside this boundary, the numbers always get smaller, and hence remain finite. Outside it they grow beyond any bound.

In the next few pages we will attempt to give you an overview of the possible forms of the basins of the finite attractor. For non-integer values of p we observe breaks in the pictures of the contour lines, which are a consequence of the way complex powers behave.

Because the calculation of powers involves complicated functions such as exponentials and logarithms, these computations take a very long time.

In each frame we see the central basin of attraction, together with contour lines for 3, 5, and 7 iterations. The attractor for $p = 1.0$, virtually a point, at first extends relatively diffusely, but by $p = 1.6$ acquires a shape which by $p = 2.0$ becomes the familiar Gingerbread Man. Between $p = 2.0$ and 3.0 it gains a further protuberance, so

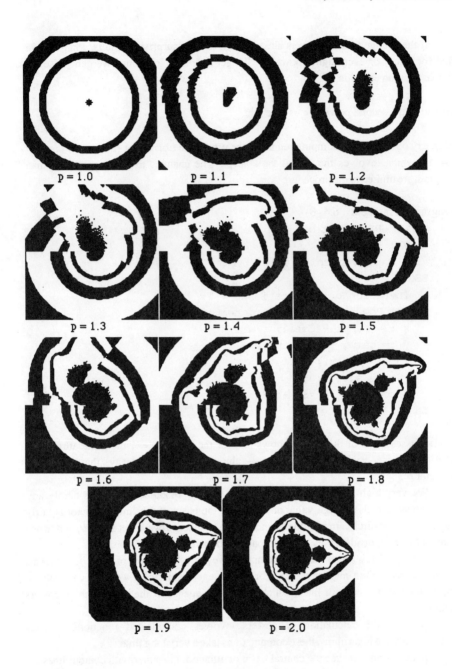

Figure 6.4–3   Generalised Mandelbrot set for powers from 1 to 2.

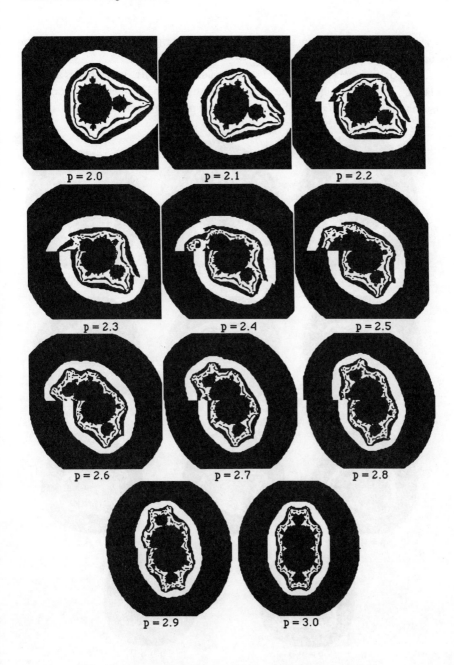

Figure 6.4-4   Generalised Mandelbrot set for powers from 2 to 3.

that we eventually obtain a very symmetric picture. The origin of the complex plane is in the middle of each frame.

The powers increase further, and at each whole number *p* a further bud is added to the basin. As already indicated, the figure grows smaller and smaller, and concentrates around the unit circle. We leave the investigation of other powers as an exercise.

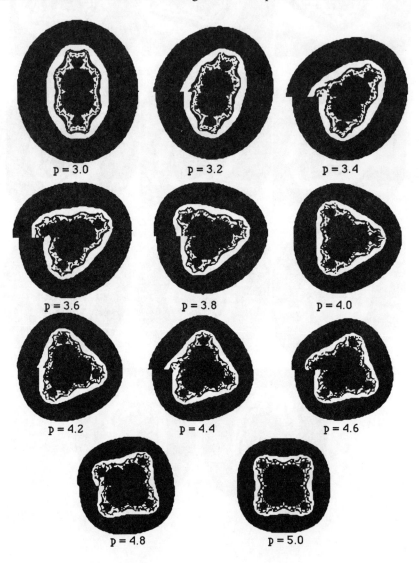

Figure 6.4–5   Generalised Mandelbrot set for powers from 3 to 5.

Computer Graphics and Exercises for §6.4

Exercise 6.4-1

Write a program to compute and draw the basin according to Curry, Garnett, and Sullivan. Either use the given procedures for complex operations, or try to formulate the algorithm one step at a time. This is not particularly easy, but has an advantage in computational speed.

Investigate the regions recommended in connection with Figure 6.4-2. An example for the region at $c = 1$ is shown in Figure 6.4-6.

Does the picture remind you of anything else you have seen in this book?

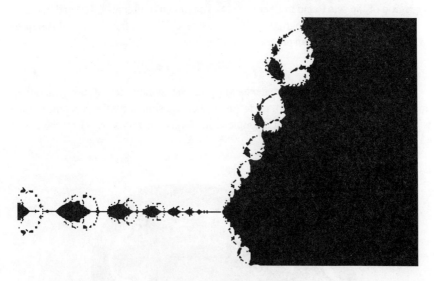

Figure 6.4-6 Section from Figure 6.4-1 near $c = 1$.

Exercise 6.4-2

Modify the program so that in the interesting region of Figure 6.4-2 a distinction is drawn between convergent and cyclic behaviour of the number sequences. Draw the figure corresponding to the Gingerbread Man. It corresponds to the numbers c for which Newton's method does not lead to a solution, but ends in a cyclic sequence.

Compare the resulting figure with the original Mandelbrot set. Spot the difference!

Exercise 6.4-3

Investigate further the elliptical regions on the edge of the 'white set' in Figure 6.4-1. Compare these. What happens to the Gingerbread Man?

Exercise 6.4–4

Develop a program to illustrate the iteration formula

$$z_{n+1} = z_n^p - c$$

graphically. Use it to investigate the symmetries of the basins for $p = 6$, $p = 7$, etc.

Try to formulate the results as a general rule.

Exercise 6.4–5

Naturally it is also possible to draw Julia sets for the iteration equation

$$z_{n+1} = z_n^p - c.$$

To get connected basins, you should probably choose a parameter c belonging to the inner region of the sets found in Exercise 6.4–4 or shown in Figures 6.4–3 and 6.4–5.

The changes to the program are fairly modest. Concentrate on the differences between the Gingerbread Man program in §6.1 and that for Julia sets in §5.2.

Exercise 6.4–6

If you have come to an understanding of the symmetries of the generalised Mandelbrot sets in Exercise 6.4–4, try to find something similar for the generalised Julia sets of the previous exercise. An example is shown in Figure 6.4–7. There the power $p = 3$ and the constant

$$c = -0.5 + 0.44*i.$$

Figure 6.4–7   Generalised Julia set.

7 New Sights – New Insights

Until now we have departed from the world of the ground-plan `Mapping` only in exceptional cases, but this chapter will show that the results of iterative calculations can be represented in other ways. The emphasis here should not be only on naked power: for the understanding of complicated relationships, different graphical methods of representation can also be used. If 'a picture is worth a thousand words', perhaps two pictures can make clear facts that cannot be expressed in words at all.

7.1   Up Hill and Down Dale

Among the most impressive achievements of computer graphics, which we encounter at every local or general election, are 3D pictures. Of course we all know that a video screen is flat, hence has only two dimensions. But by suitable choice of perspective, projection, motion, and other techniques, at least an impression of three-dimensionality can be created, as we know from cinema and television. The architectural and engineering professions employ Computer Aided Design (CAD) packages with 3D graphical input, which rapidly made an impact on television and newspapers. Although we certainly cannot compare our pictures with the products of major computer corporations of the 'Cray' class, at least we can give a few tips on how to generate pseudo-3D graphics, like those in §2.2.3.

The principle leans heavily on the mapping method of the previous chapter. The entire picture is thus divided into a series of parallel stripes. For each of them we work out the picture for a section of the 3D form. We join together the drawings of these sections, displaced upwards and to the side. We thus obtain, in a simple fashion, a 3D effect, but one without true perspective and without shadows. It also becomes apparent that there is no point in raising the iteration number too high. This of course helps to improve the computing time. To avoid searching through huge data sets, which must be checked to see which object is in front of another, we note the greatest height that occurs for each horizontal position on the screen. These points are if necessary drawn many times. For all computations concerned with iteration sequences, the iteration step should be the quantity that appears in the third direction. Two further quantities, usually the components of a complex number, form the basis of the drawing. In general we will denote these by x and y, and the third by z.

To begin with, we can generate a new pseudo-3D graphic for each picture in the previous chapters. In this type of representation Newton's method for an equation of third degree, which we know from Figure 4.3-5, generates Figure 7.1-1.

The central procedure bears the name `D3mapping`, which you will recognise because all new variables, procedures, etc. carry the prefix `D3` before their names.

The resulting picture is in a sense 'inclined' to the screen, and sticks out a bit on each side. To avoid cutting off the interesting parts of the picture, we can for instance be generous about the limits `Left` and `Right`. We have chosen a different possibility here,

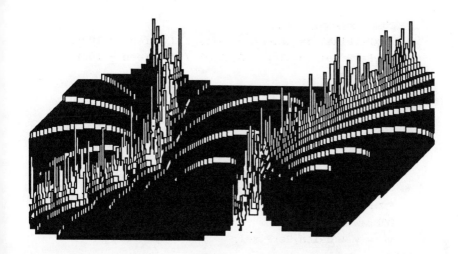

Figure 7.1-1   Boundaries between three attractors on the real axis.

in order to make this program as similar as possible to those already encountered. This gives rise to a somewhat modified denominator in the calculation of deltaxPerPixel and deltayPerPixel, and the upper limit of the FOR-loop.

The number D3factor expresses how strongly the figure is stretched in the vertical direction. The product D3factor * maximalIteration should amount to roughly a third of the screen size, that is, about 100 pixels.

The maximal coordinate of each vertical screen coordinate is stored in the array D3max. To enable this array to be passed to other procedures, a special type is defined for it. To begin with, the contents of this array are initialised with the value 0.

Program Fragment 7.1-1

```
PROCEDURE D3mapping;
    TYPE
        D3maxtype = ARRAY[0..Xscreen] OF integer;
    VAR
        D3max : D3maxtype;
        xRange, yRange, D3factor : integer;
        x, y, deltaxPerPixel, deltayPerPixel : real;

(* here some local procedures are omitted *)

BEGIN
    D3factor := 100 DIV maximalIteration;
    FOR xRange := 0 TO xScreen DO
```

```
    D3max[xRange] := 0;
deltaxPerPixel := (Right - Left) / (Xscreen - 100);
deltayPerPixel := (Top - Bottom) / (Yscreen - 100);
y := Bottom;
FOR yRange := 0 to (Yscreen - 100) DO
BEGIN
    x := Left;
    FOR xRange := 0 TO (Xscreen - 100) DO
    BEGIN
        dummy := D3ComputeAndTest (x, y, xRange, yRange);
        x := x + deltaxPerPixel;
    END;
    D3Draw (D3max);
    y := y + deltayPerPixel;
END;
END; (* D3mapping *)
```

As you see, two further procedures must be introduced: D3Draw and the functional procedure D3ComputeAndTest. The latter naturally has a lot in common with the functional procedure ComputeAndTest, which we have already met. Since the drawing is carried out in D3Draw, we must pass to it the coordinates of the currently computed point in (x,y)-space. Instead of deciding whether each individual point is to be drawn, we store the values corresponding to a given row, so that eventually we can draw the line in a single piece. D3set controls this. From the coordinates in the (x,y)-space (column, row) and the computed iteration number (height) we can work out the pseudo-3D coordinates. First the horizontal value cell is calculated, if it fits on the screen, and then the value content, which gives the vertical component. If the value is higher than the previously determined maximal value for this column of the screen, then this is inserted in its place. If it is less, then that means that it corresponds to a hidden point in the picture, and hence it is omitted.

Program Fragment 7.1-2
```
    FUNCTION D3ComputeAndTest (x, y : real; xRange, yRange :
                        integer) : boolean;
    VAR
        iterationNo : integer;
        xSq, ySq, distanceSq : real;
        finished: boolean;
    PROCEDURE startVariableInitialisation;
    (* as usual *) BEGIN END;
    PROCEDURE compute;
```

```
    (* as usual *) BEGIN END;
    PROCEDURE test
    (* as usual *) BEGIN END;


    PROCEDURE D3set (VAR D3max : D3maxType;
                     column, row, height : integer);
        VAR
            cell, content : integer;
    BEGIN
        cell := column + row - (Yscreen -100) DIV 2;
        IF (cell >= 0) AND (cell <= Xscreen) THEN
        BEGIN
            content := height * D3factor + row;
            IF content > D3max[cell] THEN
                D3max[cell] := content;
        END;
    END;   (* D3set *)


BEGIN (* D3ComputeAndTest *)
    StartVariableInitialisation;
    D3ComputeAndTest := true;
    REPEAT
        compute;
        test;
    UNTIL (iterationNo = MaximalIteration) OR finished;
    D3set (D3max, xRange, yRange, iterationNo);
END (* D3ComputeAndTest *)
```

Program Fragment 7.1-3

```
    PROCEDURE D3draw (D3max: D3maxType);
        VAR
            cell, coordinate : integer;
    BEGIN
        setPoint (0, D3max[0]);
        FOR cell := 0 TO xScreen DO
        BEGIN
            coordinate := D3max[cell];
            IF coordinate >0 THEN DrawLine (cell, coordinate);
        END;
    END;   (* D3draw *)
```

The visible parts are not drawn until an entire row has been worked out. At the

beginning of each row we start at the left side of the screen (SetPoint) and then draw
the section as a series of straight lines (DrawLine).

With a few small supplements we can retain the principle of this computation, while
improving the pictures somewhat.

The method can be changed in its first stage. In §2.2.3 we drew only every second
line, and we can do the same here. As a result the gradations are more easily
distinguished. The variable D3yStep controls the step size. For a quick survey, for
example, every tenth line suffices. The variable D3xStep runs in the oblique direction,
so that the slopes merge and the steps do not seem quite so abrupt if this value is
relatively large.

Figure 7.1-2 shows the Julia set for Newton's method applied to

$$z^3-1 = 0.$$

There every second row is drawn, so D3xStep has the value 2. Essentially, this is a
section from Figure 5.1-5, to the lower right of the middle.

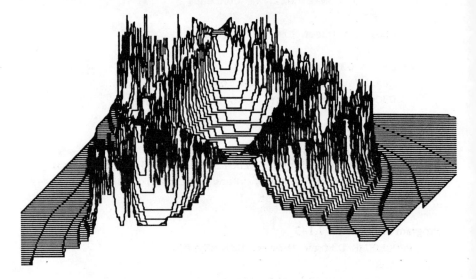

Figure 7.1-2 Julia set for Newton's method applied to $z^3-1 = 0$.

We have proceeded in fivefold steps in the next figure, Figure 7.1-3. It shows a
Julia set that you already saw in its usual fashion in Figure 5.2-5.

These large steps are useful, for example, if you just want to get a quick view of the
anticipated picture.

In order to make the central shape of the Mandelbrot set, or a Julia set, stand out
more clearly than before, it helps to draw the lines thicker – that is, to use several adjacent
lines – at these places. Our Pascal version provides a facility for doing this very easily,
using the procedure penSize. In other dialects you have to remember the beginning and

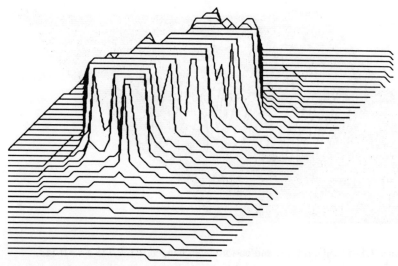

Figure 7.1-3   Julia set for $c = 0.5 + i * 0.5$.

end of this horizontal stretch and then draw a line one pixel displaced.   Figure 7.10-4 shows a Mandelbrot set drawn in this manner.

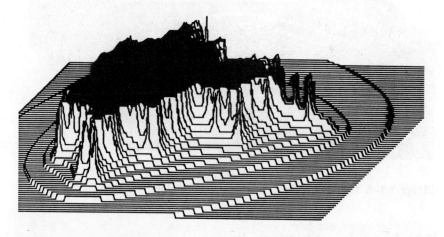

Figure 7.1-4   Gingerbread Man.

Sometimes in these pictures the slopes are interrupted by protruding peaks, and instead of mountains we find gentle valleys.   That too can be handled.   Instead of the iteration height we use the difference `MaximalIteration - iterationNo`.   In Figure 7.1-5 we see a Julia set drawn in this manner:

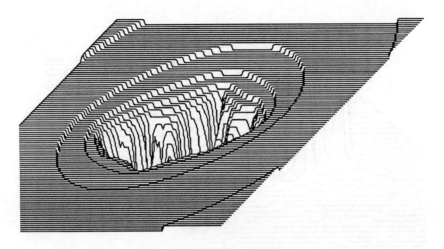

Figure 7.1-5   Julia set, top and bottom interchanged, $c = 0.745 + i * 0.113$.

Finally, as in Figure 7.1-6, we can use the reciprocal of the iteration depth in the 'third dimension', obtaining a convex imprint, but with steps of different heights.

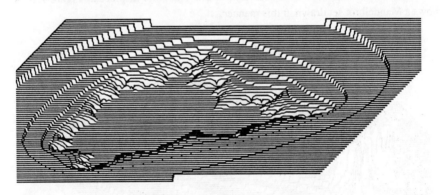

Figure 7.1-6   Gingerbread Man, inverse iteration height.

7.2   Invert It – It's Worth It!

In this section we bring to the fore something that hitherto has been at an infinite distance – namely, the attractor 'infinity'. It is clear that to do this we must forego what until now has been at the centre of things, the origin of the complex plane. The method whereby the two are interchanged is mathematically straightforward, if perhaps a little unfamiliar for complex numbers. To turn a number into another one, we invert it.

To each complex number z there corresponds another z', called its *inverse*, for which

$$z * z' = 1.$$

In other words, $z' = 1/z$, the reciprocal of z. The computational rules for complex numbers have already been set up in Chapter 4, so we can proceed at once to incorporate the idea into a Pascal program. The previous program is only changed a little. At the beginning of ComputeAndTest the appropriate complex parameter, namely z_0 for Julia sets and c for the Mandelbrot set, is inverted.

Program Fragment 7.2-1

```
FUNCTION ComputeAndTest (Creal, Cimaginary : real)
                         : boolean;

(* variables and local procedures as usual *)

    PROCEDURE invert (VAR x, y : real);
       VAR denominator : real;
    BEGIN
       denominator := sqr(x) + sqr(y);
       IF denominator = 0.0 THEN
          BEGIN
             x := 1.0E6; y := x;   {emergency solution}
          END ELSE BEGIN
             x := x / denominator;
             y := y / denominator;
          END;
       END; (* invert *)

BEGIN
    invert (Creal, Cimaginary);
    startVariableInitialisation;
    REPEAT
       compute;
       test;
    UNTIL (iterationNo = MaximalIteration) OR fisnished;
    distinguish;
END;   (* ComputeAndTest *)
```

With these changes we can recompute everything we have done so far. And look: the results are overwhelming. In Figure 7.2-1 you see what happens to the Gingerbread Man, when the underlying c-plane is inverted.

Now the Mandelbrot set appears as a black region surrounding the rest of the complex plane, which has a drop-shaped form. The buds, which previously were on the

Figure 7.2-1 Inverted Mandelbrot set.

outside, are now on the inside. The first bud, as before, is still the largest, but it is quite comparable with the remainder. And the stripes, which got wider and more solid, the further away we got from the main body? If we do not pay attention, we may miss them completely. They are all collected together in a tiny region in the middle of the picture. The middle is where the point attractor ∞ appears. The mathematical inversion has thus turned everything inside out.

Let us now compare another Gingerbread Man with his inverse. It is the one for the third power.

Figure 7.2-2 Gingerbread Man for the third power (top half; compare Figure 6.4-5).

Figure 7.2-3 Gingerbread Man for the third power, inverted (cropped on right).

And what happens to the Julia sets?   At first sight Figure 5.1-2 resembles its 'antipode' in Figure 7.2-4, but on further study the differences can be detected.

On the following pages are further pairs of normal and inverted Julia sets.

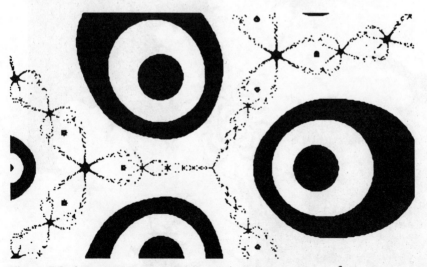

Figure 7.2-4 Inverted Julia set for Newton's method applied to $z^3 - 1 = 0$.

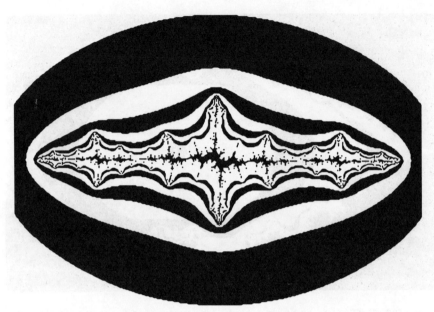

Figure 7.2-5 Julia set for $c = 1.39 - i * 0.02$.

Figure 7.2-6   Inverted Julia set for $c = 1.39 - i * 0.02$.

Figure 7.2–7 Julia set for $c = -0.35 - i * 0.004$.

Figure 7.2–8   Inverted Julia set for $c = -0.35 - i * 0.004$.

7.3 The World Is Round

It is certainly attractive to compare corresponding pictures from the previous chapter with each other, and to observe the same structures represented in different ways and/or distorted. However, from a technical point of view a large part of the pictures is superfluous, in that it contains information which can already be obtained elsewhere. A more economical picture shows just the inside of the unit circle. Everything else can be seen in the inverted picture in the unit circle. It may amount to sacrilege on aesthetic grounds, but the pure information that lies in the complex plane can be contained in two circles of radius 1. The first contains all points (x,y) for which

$$x^2+y^2 \le 1$$

and the second circle contains all the rest in inverted form:

$$x^2+y^2 \ge 1.$$

Figure 7.3-1 The entire complex plane in two unit circles.

In Figure 7.3-1 we see the Gingerbread Man in this form – of course, he has lost much of his charm. Imagine that these two circles are made of rubber: cut them out and glue the edges together back to back. Then all we need to do is blow up the picture like a balloon and the entire complex plane ends up on a sphere!

Mathematicians call this the *Riemann sphere*, in memory of the mathematician Bernhard Riemann (1826–66), who among other things made important discoveries in the area of complex numbers. His idea is explained in Figure 7.3-2.

The sphere and the plane meet at a point: the same point is the origin of the complex plane and the south pole S of the sphere. The north pole N acts as a centre of projection. A point P of the plane is to be mapped on the sphere. The connecting line NP cuts the sphere at R. The scales of the plane and the sphere are so adjusted that all points lying on the unit circle in the plane are mapped to the equator of the sphere. In the southern hemisphere we find the inner region, the neighbourhood of the origin. In the northern hemisphere we find everything that lies outside the unit circle, the 'neighbourhood of infinity'.

Why have we told you about that now? It leads naturally to new and dramatic graphical effects. Our programme for the next few pages will be to map the Gingerbread Man (playing the role of any picture in the complex plane) on to the Riemann sphere, to

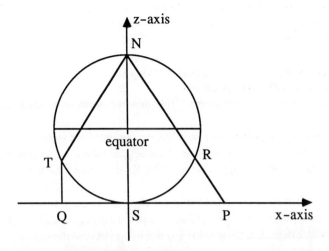

Figure 7.3-2 Projection from plane to sphere.

rotate this, and to draw the resulting picture. In order to make the resulting distortion clear, the drawing is performed by a simple orthogonal projection, which for example transforms point T to point Q in Figure 7.3-2.

For a clear and unified specification of the quantities occurring we adopt the following conventions. The final output is the complex plane, in which as before everything is computed and drawn. Each point is determined by the coordinate pair xRange, yRange (picture coordinates). The universal coordinates (for the Gingerbread Man these are c_{real} and $c_{imaginary}$) follow from these. They will be mapped on to a circle. From the two-dimensional coordinate pair a coordinate triple x_{sphere}, y_{sphere}, z_{sphere} is therefore formed.

The centre of the sphere lies above the origin of the complex plane. The x-axis and y-axis run along the corresponding axes of the plane, that is, $x_{sphere} \approx c_{real}$ and $y_{sphere} \approx c_{imaginary}$. The z_{sphere}-axis runs perpendicular to these two along the axis through the south and north poles.

We transfer all the points within the unit circle on to the southern hemisphere; all the others lie in the northern hemisphere. However, these are obscured, because we view the sphere from below. The points directly above the unit circle form the equator of the sphere. At the north pole, then, lies the point '∞'. It is in fact only one point! Two diametrically opposite points on the sphere (antipodes) are reciprocals of each other.

For mapping the complex plane we employ a second possibility as well as Riemann's; both are explained in Figure 7.3-2. There we see a section through the sphere and the plane. Suppose that a point P with coordinates (c_{real}, $c_{imaginary}$) is being mapped. The first possibility for achieving this is simply to transfer the x- and

y-coordinates, thus $x_{sphere} = c_{real}$ and $y_{sphere} = c_{imaginary}$. Then z is computed using the fact that

$$x^2 + y^2 + z^2 = 1,$$

so

$$z_{sphere} = \text{sqrt}(1.0 - \text{sqr}(x_{sphere}) - \text{sqr}(y_{sphere})).$$

This method we call *orthogonal projection*.

The second method is Riemann's. This time the north pole acts as a centre of projection.

To make different perspectives possible, we construct our figure on the Riemann sphere. Then we rotate it so that the point whose neighbourhood we wish to inspect falls at the south pole. Finally we project the resulting 3D-coordinates back into the complex plane.

On the next three pages we have again written out the procedure `Mapping`, including all local procedures. New global variables `Radius`, `Xcentre` and `Ycentre` are defined, which determine the position and size of the circle on the screen.

Because we want to use the mapping method (scan the rows and columns of the screen), we must of course work backwards through the procedure laid down above.

At the first step we test whether the point under investigation lies inside the circle. If so, we find its space coordinates using orthogonal projection. The x- and y-coordinates carry over unchanged. We compute the z-coordinate as described above. Because we are observing the sphere from below (south pole), this value is given a negative sign.

Next we rotate the entire sphere – on which each point lies – through appropriate angles (given by variables `width` and `length`). To prescribe this rotation in general we use matrix algebra, which cannot be explained further here. For further information see, for example, Newman and Sproull (1979).

From the rotated sphere we transform back into the complex plane using the Riemann method, and thus obtain the values x and y (for the Gingerbread Man these are interpreted as c_{real} and $c_{imaginary}$) for which the iteration is to be carried out.

The result of the iteration determines the colour of the original point on the screen.

Program Fragment 7.3–1

```
PROCEDURE Mapping;
  TYPE
    matrix : ARRAY [1..3,1..3] OF real;
  VAR
    xRange, yRange : integer;
    x, y, deltaxPerPixel, deltayPerPixel : real;
    xAxisMatrix, yAxisMatrix: matrix;

  PROCEDURE ConstructxRotationMatrix (VAR m: matrix; alpha
                    : real);
```

```
BEGIN
   alpha := alpha * pi / 180.0; {radian measure}
   m[1,1] := 1.0; m[1,2] := 0.0; m[1,3] := 0.0;
   m[2,1] := 0.0; m[2,2] := cos(alpha); m[2,3] :=
                  sin(alpha);
   m[3,1] := 0.0; m[3,2] := -sin(alpha); m[3,3] :=
                  cos(alpha);
END; (* ConstructxRotationMatrix *)


PROCEDURE ConstructyRotationMatrix (VAR m: matrix; beta
                  : real);
BEGIN
   beta := beta * pi / 180.0; {radian measure}
   m[1,1] := cos(beta); m[1,2] := 0.0; m[1,3] :=
                  sin(beta);
   m[2,1] := 0.0; m[2,2] := 1.0; m[2,3] := 0.0;
   m[3,1] := -sin(beta); m[3,2] := 0.0; m[3,3] :=
                  cos(beta);
END; (* ConstructyRotationMatrix *)


PROCEDURE VectorMatrixMultiply (xIn, yIn, zIn : real; m
                  : matrix;
                  VAR xOut, yOut, zOut : real);
BEGIN
   xOut := m[1,1]*xIn + m[1,2]*yIn+m[1,3]*zIn;
   yOut := m[2,1]*xIn + m[2,2]*yIn+m[2,3]*zIn;
   zOut := m[3,1]*xIn + m[3,2]*yIn+m[3,3]*zIn;
END; (* VectorMatrixMultiply *)


FUNCTION ComputeAndTest (Creal, Cimaginary : real)
                  : boolean;
   VAR
      iterationNo : integer;
      finished : boolean;
      x, y, xSq, ySq : real;


PROCEDURE StartVariableInitialisation;
BEGIN
   finished := false;
   iterationNo := 0;
   x := 0.0;
```

```
      xSq := sqr(x);
      y := 0.0;
      ySq := sqr(y);
   END (* StartVariableInitialisation *)


   PROCEDURE Compute;
   BEGIN
      y := x*y;
      y := y+y-Cimaginary;
      xSq := sqr(x);
      ySq := sqr(y);
      iterationNo := iterationNo + 1;
   END;


   PROCEDURE test;
   BEGIN
      finish := ((xSq + ySq) > 100.0);
   END;


   PROCEDURE distinguish;
   BEGIN
      ComputeAndTest := (iterationNo = MaximalIteration) OR
         (iterationNo < Bound) AND (odd(iterationNo));
   END;

BEGIN (* ComputeAndTest *)
   StartVariableInitialisation;
   REPEAT
      compute;
      test;
   UNTIL (iterationNo = MaximalIteration OR finished);
END; (* ComputeAndTest *)


FUNCTION  calculateXYok(VAR x, y : real;
                 xRange, yRange : integer) : boolean;
   VAR
      xSphere, ySphere, zSphere,
      xInter, yInter, zInter : real;
   BEGIN
      IF ((sqr(1.0 * (xRange - xCentre))+sqr(1.0*(yRange-
                  yCentre)))
            > sqr(1.0*Radius))
```

```
      THEN calculateXYok := false;
      ELSE BEGIN
         caculateXYok := true;
         xSphere := (xRange-Xcentre)/Radius;
         ySphere := (yRange-Ycentre)/Radius;
         zSphere := -sqrt(abs(1.0-sqr(xSphere)-
                    sqr(ySphere)));
         VectorMatrixMultiply
            (xSphere, ySphere, zSphere, yAxisMatrix,
               xInter, yInter, zInter);
         VectorMatrixMultiply
            (xInter, yInter, zInter, xAxisMAtrix,
               xSphere, ySphere, zSphere);
         IF zSphere = 1.0 THEN BEGIN
            x := 0.0;
            y := 0.0;
         END ELSE BEGIN
            x := xSphere/(1.0 - zSphere);
            y := ySphere/(1.0 - zSphere);
         END;
      END;
   END;   (* calculateXYok *)


BEGIN
   ConstructxRotationMatrix (xAxisMatrix, width);
   ConstructyRotationMatrix (yAxisMatrix, length);
   FOR yRange := Ycentre-Radius TO Ycentre+Radius DO
      FOR xRange := Xcentre-Radius TO Xcentre+Radius DO
      BEGIN
         IF calculateXYok (x, y, xRange, yRange) THEN
            IF ComputeAndTest (x,y) THEN
               SetPoint (xRange, yRange);
      END;
END; (* Mapping *)
```

In the first step of Mapping we set up the two rotation matrices. These are arrays of numbers which will be useful in each computation. In this way we avoid using the rather lengthy computations of sine and cosine unneccessarily often.

The main work again takes place in the two FOR loops.   You may have realised

Figure 7.3-3   Examples of Mandelbrot sets on the Riemann sphere.

that we no longer need to scan the entire screen: it is sufficient to investigate the region in which the circle appears. We therefore terminate the computation if we discover from calculateXYok that the point is outside this region. However, if we find a point inside the circle on the screen, calculateXYok computes the coordinates in the steps descibed above. The final ComputeAndTest differs only in small ways from the version already laid down.

Figure 7.3-4   Gingerbread Man rotated 60°, front and back.

Thus the important new ingredient in this chapter for representation on the Riemann sphere is the functional procedure calculateXYok. First the screen coordinates xRange and yRange, together with the variables that determine the circle, are tested, to see whether we are inside it. Do not worry about the rather curious constructions such as sqr(1.0*Radius). The variable Radius is an integer, and if for example we square 200, we may exceed the appropriate range of values for this type, which in many implementations of Pascal is limited to 32767. By multiplication by 1.0, Radius is implicitly converted to a real number, for which this limit does not hold.

The variables xSphere and ySphere, with values between -1 and +1, are computed from the screen coordinates, and from them the negative zSphere.

For the rotation we treat the three variables that define a point as a vector, which is multiplied by the previously computed matrix. The result is again a vector and contains the coordinates of the point after rotation. Intermediate values are stored in xInter, yInter, and zInter. In the next picture you once more see our favourite figure with the parameters width = 60, length = 0 (respectively 180).

7.4 Inside Story

As we have learned from our calculations, the most interesting thing about a fractal figure is undoubtedly the boundary. Its endless variety has occupied us for a great many pages. The approximation of this boundary by contour lines has led to a graphical construction of the neighbourhood of the boundary, which is often very attractive. But the inside of the basin of attraction has until now remained entirely black. That can be changed too!

In the interior the information that we can lay hands on is just the two values x and y; that is, the components of the complex number z. These are real numbers between –10 and +10. What can we do with those? First one might imagine taking one of the numbers, and using the TRUNC function to cut off everything after the decimal point, and then drawing the points for which the result is an odd number. To do this we can use the Pascal function ODD. Unfortunately the result is very disappointing, and it does not help to take $x+y$ or multiples of this. However, we have not tried everything that is possible here, and we recommend that you experiment for yourself. We have obtained the best results using a method which, to end with, we now describe. (It should be said at once that it is intended purely for graphical effect. No deeper meaning lies behind it.)

You must...

- Take the number distanceSq, which has already been found during the iteration.
- Construct its logarithm.
- Multiply by a constant insideFactor. This number should lie between 1 and 20.
- Cut off the part after the decimal point.
- If the resulting number is odd, colour the corresponding point on the screen.

All this takes place in the procedure distinguish below. The Pascal print-out has become so long that it spreads across three lines.

Program Fragment 7.4-1

```
PROCEDURE distinguish
BEGIN
    ComputeAndTest :=
        (iterationNo = MaximalIteration) AND
        ODD(TRUNC(insideFactor * ABS( ln (distanceSq))));
END;
```

Figure 7.4–1 Gingerbread Man with interior structure (`insideFactor` = 2).

Figure 7.4–2 Gingerbread Man with interior structure (`insideFactor` = 10).

The effect of this computation is to divide the interior of the basin of attraction into several parts.   Their thickness and size can be changed by altering the value of `insideFactor`.

Can you still see the Gingerbread Man in Figures 7.4-1 and 7.4-2?

Julia sets, too, provided they contain large black regions, can be improved by this method.   First we show an example without contour lines, then one with.

Figure 7.4-3 Julia set for $c = 0.5 + i*0.5$ with interior structure.

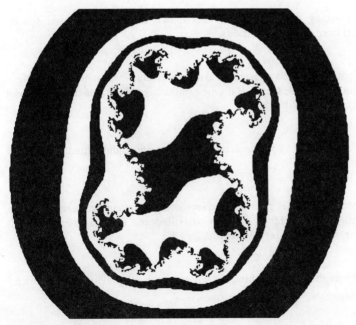

Figure 7.4-4 Julia set for $c = -0.35 + i*0.15$ with interior structure.

Computer Graphics Experiments and Exercises for Chapter 7

Exercise 7-1

Modify your programs to implement three-dimensional representation. Try out all the different methods. Parameters such as total height, or step-size in the two directions, let you generate many different pictures.

You get especially rugged contours if you allow drawing only in the horizontal and vertical directions. Then you must replace the move-to command, which generates slanting lines as well, by an appropriate instruction.

Exercise 7-2

If you wish to view the object from a different direction, you can interchange parameters, so that Left > Right and/or Bottom > Top. Alternatively or in addition you can displace the individual layers to the left instead of to the right.

Exercise 7-3

Add the inversion procedure to all programs that draw things in the complex plane. In particular you will obtain fantastic results for filigreed Julia sets.

Exercise 7-4

Further distortions occur if before or after the inversion you add a complex constant to the c- or z_0-value. In this way other regions of the complex plane can be moved to the middle of the picture.

Exercise 7-5

Transfer everything so far computed and drawn on to the Riemann sphere. Observe the objects from different directions.

Exercise 7-6

Magnified sections of the sphere are also interesting.   Not in the middle – everything stays much the same there. But a view of the edge of the planet (perhaps with another behind it) can be very dramatic.

Exercise 7-7

Perhaps it has occurred to you to draw lines of longitude and latitude? Or to try other cartographic projections?

Exercise 7.8

The entire complex plane is given a very direct interpretation in the Riemann sphere. New questions arise which, as far as we know, are still open. For example: how big is the area of the Gingerbread Man on the Riemann sphere?

Exercise 7-9

Combine the method for colouring the interior with all previously encountered pictures!

8 Fractal Computer Graphics

8.1   All Kinds of Fractal Curves

We have already encountered fractal geometric forms, such as Julia sets and the Gingerbread Man.   We will develop other aspects of their interesting and aesthetically appealing structure in this chapter.   We gradually leave the world of dynamical systems, which until now has played a leading role in the formation of Feigenbaum diagrams, Julia sets, and the Gingerbread Man.   There exist other mathematical functions with fractal properties.   In particular we can imagine quite different functions, which have absolutely nothing to do with the previous background of dynamical systems.   In this chapter we look at purely geometric forms, which have only one purpose - to produce interesting computer graphics.   Whether they are more beautiful than the Gingerbread Man is a matter of personal taste.

Perhaps you already know about the two most common fractal curves. The typical structure of the Hilbert and Sierpiński curves is shown in Figures 8.1-1 and 8.1-2. The curves are here superimposed several times.

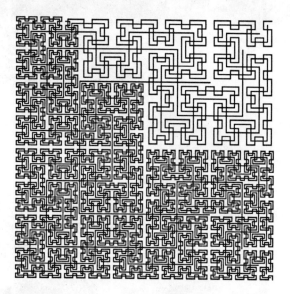

Figure 8.1-1   'Genesis' of the Hilbert curve.

As with all computer graphics that we have so far drawn, the pictures are 'static' representations of a single situation at some moment.   This is conveyed by the word 'genesis'. Depending on the parameter n, the number of wiggles, the two 'space-filling' curves become ever more dense. These figures are so well known that in many computer science books they are used as examples of recursive functions.   The formulas for computing them, or even the programs for drawing them, are written down there: see for example Wirth (1983). We therefore do not include a detailed description of these two curves. Of course, we encourage you to draw the Hilbert and Sierpiński curves, even though they are well known.

Figure 8.1-2 'Genesis' of the Sierpiński curve.

Recursive functions are such that their definition is in some sense 'part of itself'. Either the procedure calls itself (direct recursion), or it calls another, in which it is itself required (indirect recursion). We refer to recursion in a graphical representation as self-similarity.

We obtain further interesting graphics if we experiment with *dragon curves* or *C-curves*. Figures 8.1-3 to 8.1-5 show the basic form of these figures.

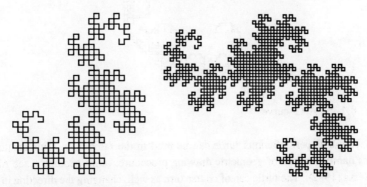

Figure 8.1-3 Different dragon curves.

Before you proceed to your own experiments, with the aid of the exercises, we will provide you with some drawing instructions in the form of a program fragment. Having set up this scheme you can easily experiment with fractal figures.

It is simplest if we start the drawing-pen at a fixed position on the screen and begin drawing from there. We can move forwards or backwards, thereby drawing a line.

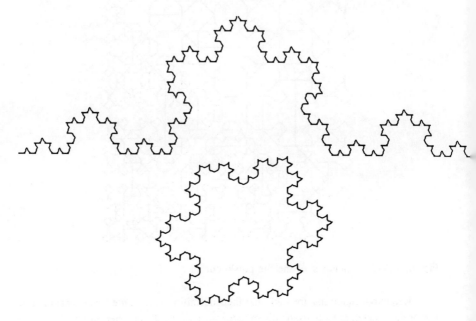

Figure 8.1-4   Different Koch curves.

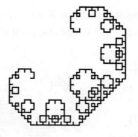

Figure 8.1-5   A C-curve.

Seymour Papert's famous *turtle* can be used in the same way. This creature has given its name to a type of geometric drawing procedure: *turtle geometry*: see Abelson and diSessa (1982). The turtle can of course turn as well, changing the direction in which it moves. If the turtle is initially placed facing the x-direction, the command

```
     forward (10)
```
means that it takes 10 steps in the direction of the positive *x*-axis. The command
```
     turn (90)
```
rotates it clockwise through 90°. A repeat of the command
```
     forward (10)
```
makes the turtle move 10 steps in the new direction. Basically, that is everything that the turtle can do. On some computers, turtle graphics is already implemented. Find out more from your handbook.

We have also formulated the main commands in Pascal, so that you can easily carry out this process on your computer. This scheme, which works with 'relative' coordinates, is as follows:

Program Fragment 8.1-1 (Procedures for turtle graphics)

```
     PROCEDURE forward (step: integer);
        CONST
           pi = 3.1415;
        VAR
           xstep, ystep: real;
     BEGIN
        xstep := step * cos ((turtleangle*pi)/180);
        ystep := step * sin ((turtleangle*pi)/180);
        turtlex := turtlex + trunc(xstep);
        turtley := turtley + trunc(ystep);
        drawLine (turtlex, turtley);
     END;


     PROCEDURE backward (step: integer);
     BEGIN
        forward (-step);
     END;


     PROCEDURE  turn (alpha: integer);
     BEGIN
        turtleangle := (turtleangle + alpha) MOD 360;
     END;


     PROCEDURE startTurtle;
     BEGIN
        turtleangle := 90; turtlex := startx; turtley := starty;
        setPoint (startx, starty);
     END;
```

With this scheme you are in a position to carry out the following exercises. We hope you have a lot of fun experimenting.

Computer Graphics Experiments and Exercises for §8.1

Exercise 8.1-1

Set up a program to draw the Hilbert graphic. Experiment with the parameters. Draw pictures showing overlapping Hilbert curves. Draw the Hilbert curve tilted to different inclinations.   Because the task is easier with a recursive procedure, we will describe the important features now. The drawing instructions for the Hilbert curve are:

```
PROCEDURE hilbert (depth, side, direction : integer);
BEGIN
   IF depth > 0 THEN
   BEGIN
      turn (-direction * 90);
      hilbert (depth-1, side, -direction);
      forward (side);
      turn (direction * 90);
      hilbert (depth-1, side, direction);
      forward (side);
      hilbert (depth-1, side, direction);
      turn (direction * 90);
      forward (side);
      hilbert (depth-1, side, direction);
      turn (direction * 90);
   END;
END;
```

As you have already discovered in earlier pictures, you can produce different computer-graphical effects by varying the depth of recursion or the length of the sides. The value for the direction of the Hilbert curve is either 1 or -1.

Exercise 8.1-2

Experiment with the following curve. The data are, e.g.:

depth	side	startx	starty
9	5	50	50
10	4	40	40
11	3	150	30
12	2	150	50
13	1	160	90
15	1	90	150

The drawing instructions for the dragon curve are:

```
PROCEDURE dragon (depth, side: integer);
BEGIN
    IF depth > 0 THEN
        foward (side)
    ELSE IF depth > 0 THEN
        BEGIN
            dragon (depth-1, trunc (side));
            turn (90);
            dragon (-(depth-1), trunc (side));
        END
    ELSE
        BEGIN
            dragon (-(depth+1),trunc (side));
            turn (270);
            dragon (depth+1,trunc (side);
        END;
END;
```

Exercise 8.1-3

Experiment in the same way with the Koch curve. The data are, e.g.:

depth	side	startx	starty
4	500	1	180
5	500	1	180
6	1000	1	180

The drawing procedure is:

```
PROCEDURE koch (depth, side: integer);
BEGIN
    IF depth = 0 then forward (side)
    ELSE BEGIN
        koch (depth-1, trunc (side/3)); turn (-60);
        koch (depth-1, trunc (side/3)); turn (120);
        koch (depth-1, trunc (side/3)); turn (-60);
        koch (depth-1, trunc (side/3));
    END;
END;
```

Exercise 8.1-4

In the lower part of Figure 8.1-4 you will find the snowflake curve. Snowflakes can be built from Koch curves. Develop a program for snowflakes and experiment with it:

```
PROCEDURE snowflake;
BEGIN
   koch (depth, side); turn (120);
   koch (depth, side); turn (120);
   koch (depth, side); turn (120);
END;
```

Exercise 8.1-5

Experiment with right-angled Koch curves. The generating method and the data for pictures are, e.g.:

```
PROCEDURE rightkoch (depth, side: integer);
BEGIN (* depth = 5, side = 500, startx = 1, starty = 180 *)
   IF depth = 0 THEN forward (side)
   ELSE
   BEGIN
      rightkoch (depth-1, trunc (side/3)); turn (-90);
      rightkoch (depth-1, trunc (side/3)); turn (90);
      rightkoch (depth-1, trunc (side/3)); turn (90);
      rightkoch (depth-1, trunc (side/3)); turn (-90);
      rightkoch (depth-1, trunc (side/3));
   END;
END;
```

Exercise 8.1-6

Experiment with different angles, recursion depths, and side lengths. It is easy to change the angle data in the procedural description of C-curves:

```
PROCEDURE cCurve (depth, side : integer);
BEGIN (* depth = 9,12; side = 3; startx = 50, 150;
                     starty = 50, 45 *)
   IF depth = 0 THEN forward (side)
   ELSE
   BEGIN
      cCurve (depth-1, trunc (side)); turn (90);
      cCurve (depth-1, trunc (side)); turn (-90);
   END;
END;
```

Exercise 8.1-7

At the end of the C-curve procedure, after the turn(-90), add yet another procedure call cCurve (depth-1, side). Experiment with this new program. Insert the new statement after the procedure call turn(90).

Experiment also with C-curves by changing the side length inside the procedure, for example forward (side/2), etc.

Exercise   8.1-8

Draw a tree-graphic. Use the following general scheme:

```
PROCEDURE tree (depth, side: integer);
BEGIN (* e.g. depth = 5, side = 50 *)
   IF   depth > 0 THEN
   BEGIN
      turn (-45); forward (side);
      tree (depth-1, trunc (side/2)); backward (side);
               turn (90);
      forward (side);
      tree (depth-1, trunc (side/2)); backward (side);
               turn (-45);
   END;
END;
```

We have collected together the solutions to these exercises in §11.2 in a complete program. Look there if you do not wish to develop the programs for yourself.

If you have done Exercise 8.1-10, the picture will doubtless have reminded you of structures in our natural surroundings - hence the name. We thereby open up an entire new chapter of computer graphics. Computer graphics experts throughout the world have been trying to construct convincing natural forms. The pictures that emerge from the computer are landscapes with trees, grass, mountains, clouds, and lakes. Of course you need rather fancy programs to draw really convincing graphics.

But even with small computers we can produce nice things.

8.2   Landscapes: Trees, Grass, Clouds, Mountains, and Lakes

Since 1980, when the discovery of the Mandelbrot set opened a new chapter in fundamental mathematical research, new discoveries in the general area of fractal structure have occurred almost daily. Examples include fractal models of cloud formation and rainfall in meteorology. Several international conferences on computer graphics have been devoted to this; see SIGGRAPH (1985).

Likewise, such procedures are part of the computer-graphical cookery used for special effects in films. The latest research objective is the convincingly natural representation of landscapes, trees, grass, and clouds. Some results are already available in the products of the American company LucasFilms, which in recent years has made several well-known science fiction films. It has its own team for basic scientific research into computer graphics. Not surprisingly, conference proceedings sometimes bear the address of the LucasFilms studios, e.g.Smith (1984). It is now only a matter of time and computing power before films contain lengthy sequences calculated by a computer.

Some examples show how easy it is to construct grasses and clouds on a personal computer.

Figure 8.2–1 Grass and twigs.

Figure 8.2-1 shows a tree curve (see exercises for §8.1). The generating procedure for Figure 8.2-2 is already known. Namely, it is a dragon curve of high recursion depth and side length 1. In fact, with all fractal figures, you can experiment, combining the recursive generating procedures together or generalising them. Changing the depth, side length, or angles according to parameters can produce baffling results, which it is quite likely that nobody has ever seen before. To simulate natural structures such as grass, trees, mountains, and clouds, you should start with geometrical figures whose basic form resembles them, and fit them together.

A new possibility arises if we change the parameters using a random number generator.

For example we can make the angle change randomly between 10 and 20 degrees, or the length. You can of course apply random numbers to all of our other figures. We recommend that you first become very familiar with the structure, to get the best possible grasp of the effect of parameter changes. Then you can introduce random numbers effectively.

Figure 8.2-3 shows examples of such experiments. For example, the command `forward (side)` may be changed by a random factor. Thus the expression `side` is replaced by `side * random(10,20)`. This is what we have done to get the left and middle pictures in Figure 8.2-3. More natural effects are obtained if we make only small changes. Thus in the right-hand picture the expression `side` is replaced by `side*random (side, side+5)`. The `random` function always produces values

Figure 8.2-2 Two-dimensional cloud formation.

between the two bracketed bounds (compare Exercise 8.2-2).

Of course the different pictures in Figure 8.2-3 are not easily reproducible. We wrote a program to draw an endless sequence of such grass-structures, and selected interesting pictures from the results. You can collect your best grass- and cloud-structures on an external storage medium (optical, hard, or floppy disk).

For experiments with grass-structures, use the above description (see Exercise 8.2-2) as the basis for your own discoveries. The pictures were generated with the fixed value depth = 7, angle = 20 or 10, and random side lengths.

Modern computers sometimes have the facility to merge or combine parts of pictures with others. It is also possible, using drawing programs such as MacPaint (Macintosh) or special programs on MS-DOS or Unix machines, to mix parts of pictures and work on them further.

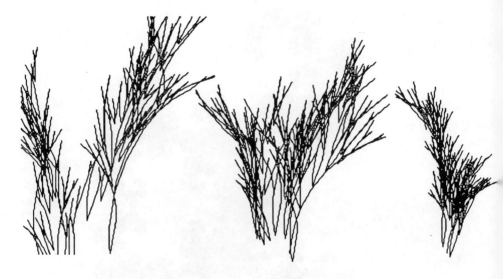

Figure 8.2-3 Different grasses.

Having obtained the many small units that arise in nature, we wish to put them together into a complete landscape. The principle for generating landscapes was explained in the Computer Recreations column of *Scientific American*, see Dewdney (1986a).

Start with a triangle in space. The midpoints of the three sides are displaced upwards or downwards according to a random method. If these three new points are joined, we have in general four different triangles, with their own particular sides and corners. The method is carried out in turn for the resulting small triangles. To obtain a realistic effect, the displacment of the midpoints should not be as great as the first time. From 1 triangle with 3 vertices, we thus obtain 4 triangles with 6 vertices, 16 triangles with 15 vertices, 64 triangles with 45 vertices. After 6 iterations we have 4096 tiny triangles contained by 2145 vertices. Now we draw these. As you see in Figure 8.2-4, the result is in fact reminiscent of mountains. For reasons of visibility we have drawn only two of the three sides of each triangle. Also, we have supplemented the picture to form a rectangle.

We avoid describing a program here: instead we have given an example of a complete program in §11.3. Because the computation of the four points takes quite a long time, we have only specified the displacements. So, for example, with the same data-set you can draw a sunken or a raised landscape. Or you can ignore everything below some particular value of the displacment and draw just single dots. In this way an impression of lakes and seas is created, as in Figure 8.2-4. In fact these pictures are already quite remarkable. Of course there are many other gadgets in the LucasFilm

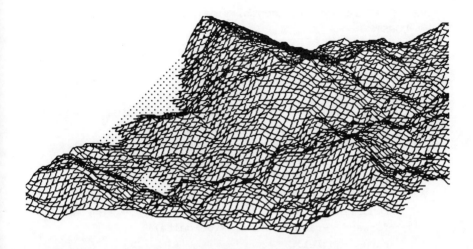

Figure 8.2–4 Fractal landscape with lake and mountains.

studio's box of tricks. Even the computer–graphical methods are improved from day to day, so that fiction and reality merge smoothly into each other. But that is another story.

Computer Graphics Experiments and Exercises for §8.2

Exercise 8.2–1

With the aid of the above description, develop a program to represent twigs:

```
PROCEDURE twig (depth, side, angle : integer);
BEGIN
    IF depth > 0 THEN
    BEGIN
        turn (-angle);
        forward (2*side);
        twig (depth-1, side, angle);
        backward (2*side); turn (2*angle);
        forward (side);
        twig (depth-1, side, angle);
        backward (side); turn (-angle);
    END;
END;
```

Exercise 8.2.–2

Experiment with grass structures too:

```
PROCEDURE randomTwig (depth, side, angle: integer);
  CONST
      delta = 5;
BEGIN
  IF depth > 0 THEN
  BEGIN
    turn (-angle);
    forward (2 * random (side, side+delta));
    randomTwig (depth-1, side, angle);
    forward(-2 * random (side, side+delta));
    turn (2*angle);
    forward (random (side, side+delta));
    randomTwig (depth-1, side, angle);
    forward (-random (side, side+delta));
    turn (-angle);
  END;
END;
```

As data for the two exercises, use $7 \leq$ depth ≤ 12, $10 \leq$ side ≤ 20, $10 \leq$ angle ≤ 20.

Try to implement a suitable procedure random.

If you modify the program descriptions in other ways, you can generate still more pictures.

8.3   Graftals

Besides fractal structures, nowadays people in the LucasFilm studios or university computer graphics laboratories also do experiments with *graftals*. These are mathematical structures which can model much more professionally the things we drew in the previous section: plants and trees. Like fractals, graftals are characterised by self-similarity and great richness of form under small changes of parameters. For graftals, however, there is no mathematical formula such as those we have found for the simple fractal structures we have previously investigated. The prescription for generating graftals is given by so-called *production rules*. This concept comes from information theory. The grammatical structure of progamming languages is defined by production rules. With production rules it is possible to express how a language is built up. We use something similar to model natural structures, when we wish to lay down formal rules for the way they are constructed.

The words of our language for generating graftals are strings formed from the symbols '0', '1', and square brackets '[', ']'. For instance, the string 01[11[01]] represents a graftal.

A production rule (substitution rule) might resemble the following:

   $0 \rightarrow 1[0]1[0]0$
   $1 \rightarrow 11$

[→ [
] →]

The rule given here is only an example. The rule expresses the fact that the string to the left of the arrow can be replaced by the string to its right. If we apply this rule to the string 1[0]1[0]0, we get

11[1[0]1[0]0]11[1[0]1[0]0]1[0]1[0]0.

Another rule (1111[11]11[111]1) represents, for instance, a tree or part of a tree with a straight segment 7 units long. Each 1 counts 1 unit, and in particular the numbers in brackets each represent a twig of the stated length. Thus the first open bracket represents a twig of length 2. It begins after the first 4 units of the main stem. The 4 units of the main stem end at the first open bracket. The main stem grows a further 2 units. Then there is a twig of length 3, issuing from the 6th unit of the main stem. The main stem is then finally extended by 1 unit (Figure 8.3-1).

This example shows how by using and interpreting a particular notation, namely 1111[11]11[111]1, a simple tree structure can be generated. Our grammar here has the alphabet {1,[,]}. The simplest graftal has the alphabet {0,1,[,]}. The '1' and '0' characterise segments of the structure. The brackets represent twigs.

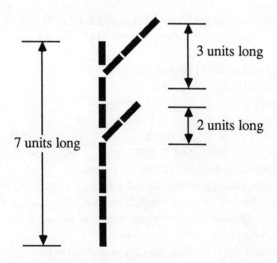

3 units long

2 units long

7 units long

Figure 8.3-1 The structure 1111[11]11[111]1.

Once we know how to interpret these bracketed structures, all we require is a method for producing structures with many twigs by systematically applying production rules. This variation on the idea just explained can be obtained as follows:

We agree that each part of the structure is generated from a triple of zeros and ones, such as 101 or 111. Each triple stands for a binary number, from which, with the aid of the appropriate production rule, the graftal is constructed.

For example, '101' expresses in binary the number 5, and '111' the number 7.

We provide a table of binary numbers to help you remember:

Decimal	Binary
0	000
1	001
2	010
3	011
4	100
5	101
6	110
7	111

Table 8.3-1 Decimal and binary numbers (0 – 7).

We express the production rule for building the graftal in terms of such binary numbers. Because there are 2^3 possible combinations of digits $\{0,1\}$, we must specify 8 production rules.

An example of such an 8-fold production rule is:

0.1.0.1.0.00[01]0.0

or, otherwise written,

position:	0	1	2	3	4	5	6	7
rule:	**0.**	**1.**	**0.**	**1.**	**0.**	**00[01].**	**0.**	**0**
binary number:	000	001	010	011	100	101	110	111

An example will explain how the construction is carried out.

We take each individual digit of the rule and generate the binary triple. Compare the triple with the production rule. Replace the digit with the corresponding sequence from the rule. To do this, the formation of the triples must also be governed by a rule. This takes (e.g.) the following form:

For a single zero or one at the beginning, a 1 is added to left and right. For a pair of numbers, first a 1 is added to the left and then the pair is repeated with a 1 added to the right.

Start with a 1. We add 1 to left and right: this gives 111. If we had started with 0 we would have got 101.

Our production rule consists mainly of the strings '0' or '00[01]'. A '1' generates '0' by applying the rule, a '0' generates '00[01]'.

Applying the rule to these strings then leads to complicated forms.

We follow the development through several generations, beginning with '1':

Generation   0

| 1 | → transformation → | 111 |
| 111 | → rule → | 0 |

(An isolated '1' at the start is extended to left and right by '1'.)

Generation   1

| 0 | → transformation → | 101 |
| 101 | → rule → | 00[01] |

(An isolated '0' at the start is extended to left and right by '1'.)

Generation   2

| 00[01] | → transformation→ | 100 001 [001 011] |
| 100 001 [001 011] → rule → | | 01[11] |

(For pairs of digits 1 is added to the left and to the right.)

Generation   3

| 01[11] | → transformation → | 101 011[111 111] |
| 101 011[111 111] → rule → | | 00[01] 1 [0 0] |

(If the main track is broken by a branch, the last element of the main branch is used for pair-formation.)

Generation   4

| 00[01] 1 [0 0] | → transformation → | 100 001 [001 011] 011 [100 001] |
| 100 001 [001 011] 011 [100 001]   → rule → | | 01[11]1[01] |

(if the main track is broken by a branch, the last element of the main branch is used for pair-formation.)

Generation   5

| 01[11]1[01] | → transformation → | 101 011 [111 111] 111 [101 011] |
| 101 011 [111 111] 111 [101 011]   → rule → | 00[01] 1 [0 0] 0 [00[01] 1] |

Generation   6

would then be 01[11]0[01]00[01][01[11]1].

In order to make the above clearer, let us consider generations 3 and 4 (see Figure 8.3-2).

Generation 3 comprises 01[11].

   Take the pair 01 and extend left by 1 to get 101.
   Take the pair 01 and extend right by 1 to get 011.
   The brackets follow.
   Take the pair 11 and extend left by 1 to get 111.
   Take the pair 11 and extend right by 1 to get 111.

Generation 4 comprises 00[01]1[01]. At generation 4 there is a difficulty. The main branch (001) is broken by a side–branch (brackets): 00[01]1[01].

   Take the pair 00 and extend left by 1 to get 100.
   Take the pair 00 and extend right by 1 to get 001.
   Take the pair 01 and extend left by 1 to get 100.
   Take the pair 01 and extend right by 1 to get 011.

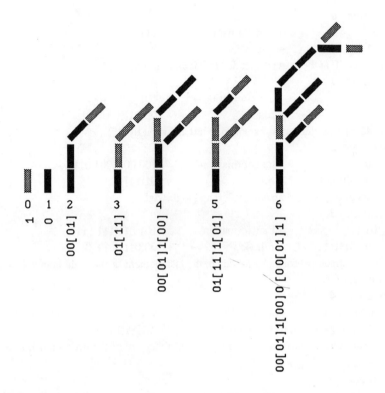

Figure 8.3-2   Construction of a graftal.

At the next digit 1 we must attach the previous digit 0 from the main branch. If we do this we get the pair 01. Now we have:

        Take the 01 and extend right by 1 to get 011.

        The two remaining digits [01] then give 101 and 011.

Figure 8.3-2 shows how this graftal builds up over 6 generations.

Now that this example has clarifed the principles for constructing graftals, we will exhibit some pictures of such structures.

In Figure 8.3-4 we show the development of a graftal from the 4th to the 12th generation. A sequence up to the 13th generation is shown in Figure 8.3-5.

We should point out that the development of graftal structures takes a lot of computation - hence time. In fact, the later generations can tie up your computer for an entire night. Your computer may also exceed the available RAM storage capacity (about 1 MB), so that generations this high cannot be computed (see the description in §11.3).

Figure 8.3-3 Graftal-plant.

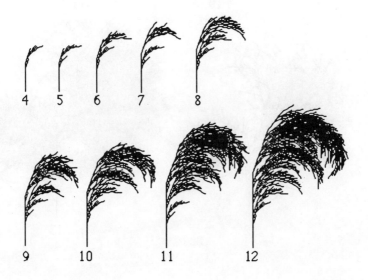

Figure 8.3-4 Development of a graftal from the 4th to the 12th generation.

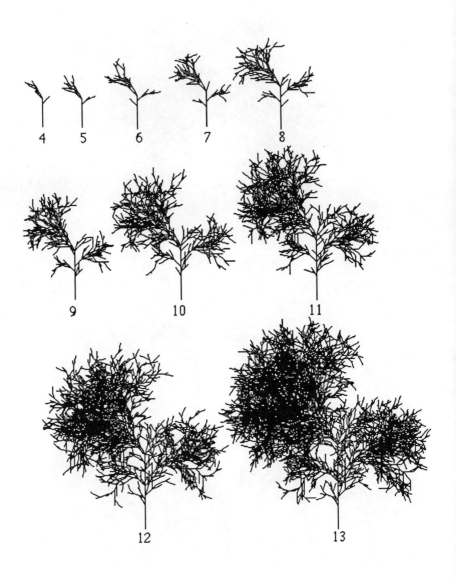

Figure 8.3-5 A graftal from the 4th to the 13th generation.

Computer Graphics Experiments and Exercises for §8.3

The experimental field of graftals is still not widely known. The following examples will get you started. A program for graphical representation is given in §11.3.[1]

Exercise 8.3-1
Experiment with graftals of the following structure:
Rule:0.1.0.11.[01].0.00[01].0.0
Angle: -40,40,-30,30
Number of generations: 10.

Exercise 8.3-2
Experiment with graftals of the following structure:
Rule: 0.1.0.1.0.10 [11].0.0
Angle: -30,20,-20,10
Number of generations: 10.

Exercise 8.3-3
Experiment with graftals of the following structure:
Rule:0.1[1].1.1.0.11.1.0
Angle: -30,30,-15,15,-5,5
Number of generations: 10.

Exercise 8.3-4
Experiment with graftals of the following structure:
Rule:0.1[1].1.1.0.11.1.0
Angle: -30,30,-20,20
Number of generations: 10.

Exercise 8.3-5
Experiment with graftals of the following structure:
Rule:0.1[01].1.1.0.00[01].1.0
Angle: -45,45,-30,20
Number of generations: 10.

Exercise 8.3-6
Vary the above examples in any way you choose, changing the production rule, the angle, or the number of generations.

---

[1] The idea of graftals has been known for some time in the technical literature; see Smith (1984), SIGGRAPH (1985). We first thought about carrying out this type of experiment on a PC after reading a beautiful essay about them. In this section we have oriented ourselves following the examples in that article: Estvanik (1986), p. 46.

8.4   Repetitive Designs

Now things get repetitive.   What in the case of graftals required the endless application of production rules, resulting in ever finer structure, can also be generated by other – simpler – rules.   The topic in this section is computer–graphical structures which can be continued indefinitely as 'repetitive designs' – rather like carpets.   The generating rules are not production rules, but algorithms constructed in the simplest manner.   The structures that are generated are neither fractals nor graftals, but 'repetitals', if you wish.

We found the simple algorithms in the Computer Recreations column of *Scientific American*; see Dewdney (1986b).   Naturally we immediately began to experiment.   The pictures we produced will not be concealed any longer:

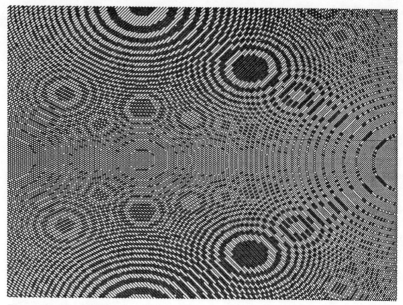

Data: 0, 10, 0, 20

Figure 8.4-1 Interference pattern 1.

The program listing for Figures 8.4-1 to 8.4-3 is very simple:

Program Fragment 8.4-1

```
PROCEDURE Conett;
    VAR
        i, j, c: integer;
        x, y, z: real;
    BEGIN
```

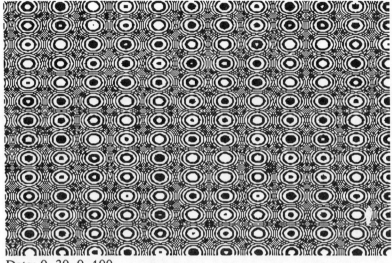

Data: 0, 30, 0, 100

Figure 8.4-2 Interference-pattern 2

Data: 0, 50, 0, 80

Figure 8.4-3 Interference-pattern 3.

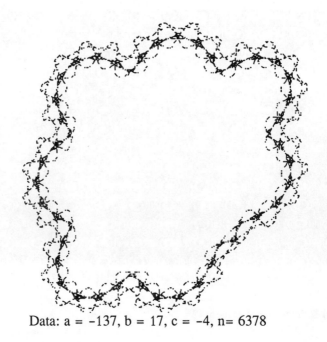

Data: a = −137, b = 17, c = −4, n= 6378

Figure 8.4-4 Garland.

```
FOR i := 1 TO Xscreen DO
   FOR j := 1 TO Yscreen DO
   BEGIN
      x := Left + (Right-Left)*i/Xscreen;
      y := Bottom + (Top - Bottom) * j / Yscreen;
      z := sqr(x) + sqr(y);
      IF trunc (z) < maxInt THEN
      BEGIN
         c := trunc (z);
         IF NOT odd (c) THEN SetPoint (i,j);
      END;
   END;
END;
```

Input data for Left, Right, Bottom, Top are given in the figures. Again the richness of form obtained by varying parameters is astonishing. The idea for this simple algorithm is due to John E. Conett of the University of Minnesota; see Dewdney (1986b).

A quite different form of design can be obtained with Barry Martin's algorithm (Figures 8.4-4ff.). Barry Martin, of Aston University in Birmingham, devised a method which is just as simple as the above method of John Conett. It depends on two simple formulas, which combine together the sign, absolute value, and root functions. The sign function has the value +1 or –1, depending on whether the argument x is positive or negative. If $x = 0$ then the sign function equals 0.

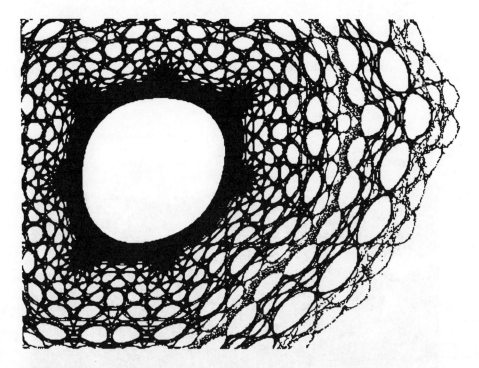

Figure 8.4–5   Spiderweb with $a = -137$, $b = 17$, $c = -4$, $n = 1\ 898\ 687$.

The program listing for Figs 8.4-4ff. is as follows:

Program Fragment 8.4–2

```
FUNCTION sign (x: real) : integer;
BEGIN
    sign := 0;
    IF x <> 0 THEN
        IF x < 0 THEN
            sign := -1
        ELSE IF x > 0 THEN
            sign := 1;
END;
```

```
PROCEDURE Martin1;
   VAR
      i,j : integer;
      xOld, yOld, xNew, yNew : real;
BEGIN
   xOld := 0;
   yOld := 0;
   REPEAT
      SetUniversalPoint (xOld, yOld);
      xNew := yOld - sign (xOld)*sqrt (abs (b*xOld-c));
      yNew := a - xOld;
      xOld := xNew;
      yOld := yNew;
   UNTILButton;
END;
```

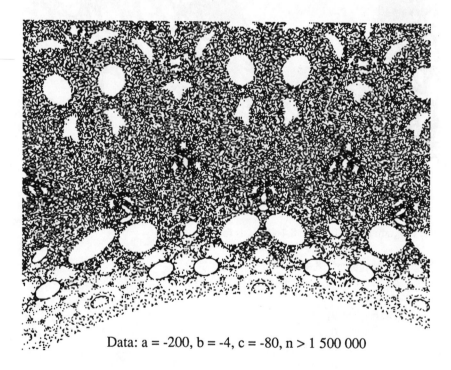

Data: a = -200, b = -4, c = -80, n > 1 500 000

Figure 8.4-6 Cell culture.

With these completely different pictures we bring to and end our computer-graphical experiments, and once more recommend that you experiment for yourself.

In previous chapters you have been confronted with many new concepts and meanings. It all began with experimental mathematics and measles. We reached a provisional end with fractal computer graphics and now with carpet designs. Other aspects will not be discussed further.

Until now we have made no attempt to structure our discoveries in this new science, on the border between experimental mathematics and computer graphics. We attempt this in the next chapter under the title 'step by step into chaos'. After a glance back to the 'land of infinite structures', our investigations of the relation between order and chaos will then come to an end.

In the chapters after these (Chapter 11 onwards), we turn to the solutions to the exercises, as well as giving tricks and tips which are useful for specific practical implementations on various computer systems.

Computer Graphics Experiments and Exercises for §8.4

Exercise 8.4-1

Implement program listing 8.4-1 and experiment with different data in the range [0,100] for the input size.

Try to find how the picture varies with parameters.

Which parameters produce which effects?

Exercise 8.4-2

Implement program listing 8.4-2 and experiment with different data for the input variables a, b, c.

Try to find how the picture varies with parameters.

Which parameters produce which effects?

Exercise 8.4-3

In program listing 8.4-2 replace the statement

```
xNew := yOld - sign (xOld)*sqrt (abs (b*xOld-c));
```
by
```
xNew := yOld - sin (x);
```
and experiment (as in Exercise 8.4-2).

9 Step by Step into Chaos

The people of Bremen are used to all kinds of weather. But what happened in the summer of 1985 in Bremen exceeded the proverbial composure of most inhabitants. On the 24th of July the telephone in the weather office was running hot. Angry callers complained about the weather forecast which they had read in the morning in the *Weserkurier*. There, a day later, you can read a lengthy article on the summer weather:

> 'Lottery players discover twice a week whether they have landed a jackpot or a miserable failure. The game of Bremen weather can now be played every day. For example, whoever took yesterday's forecast from the Bremen weather office as the basis of his bet may as well have thrown it down the drain. Instead of "heavy cloud and intermittent rain with temperatures of 19°" we got a beautiful sunny summer's day with blue skies, real bikini weather.'

What happened to make the meteorologists go so wrong, and what does it have to do with complex systems and the concept of chaos?

The starting point for any weather forecasts is the *synopsis*, that is, a survey of the current weather over a large area. The basis for this is measurements and observations from weather stations on land or at sea. For example, every day radiosondes climb into the atmosphere, and during their climb to almost 40 km in altitude they measure temperature, humidity, and pressure. A whole series of parameters such as pressure, temperature, dewpoint, humidity, cloud cover and windspeed are collected in this manner and input into a forecasting model, which is calculated with the aid of high-speed supercomputers.

Such a forecasting model is a system of differential equations, which describes a closed, complex system. It is derived from particular physical assumptions, which you may perhaps have encountered during your schooldays. Within it are quantities such as:

* impulse
* mass (continuity equation)
* humidity (balance equation for specific heat)
* energy (primary concept in the study of heat)

As already touched upon in the first chapter, the German meteorological department uses a lattice-point model, which lays down a 254 km mesh (at 60° N) over the northern hemisphere. It recognises 9 different heights, so there are about $5000 \times 9 = 45\,000$ lattice points. Influences from ground level (soil, ocean) and topography (mountains, valleys) are taken into consideration at individual lattice points.

For each lattice point the evolution of the atmospheric state is computed by 'mathematical feedback' from the initial state. Every 3.33 minutes the current state of a lattice point is taken as the initial value (input) for a new computation. Such a time-step takes the computer 4 seconds in 'real time'. The preparation of a 24-hour forecast takes about 30 minutes.

The model equations contain the following parameters: temperature, humidity, wind, pressure, cloud cover, radiation balance, temperature and water content of the ground, water content of snow cover, warm currents at ground level, and surface temperature of the ocean.

This is an impressive number of lattice points and parameters, which argues for the complexity, and equally for the realistic nature, of the forecasting model.

But we must be more honest and consider how likely it is, in spite of this, that a false prediction might be made. There is a whole series of evident possibilities for failure in the model:

- Uncertainty and inaccuracy in the analysis, due to poverty of data (e.g. over the ocean) or inadequate description of the topography.

- Space-time solution of the weather parameters in the prediction model. Of course, the finer the lattice and the shorter the time-step, the better the prediction will be. But too small units lead to long computation times!

- Various processes in the atmosphere are only understood empirically: that is, they are not studied through physically grounded equations, but with the help of parameters obtained in experiments. Thus convection, precipitation regions, ground processes, and the interaction of ground and atmosphere are all described empirically.

- Various boundary conditions cannot be represented well. Among these are the influence of 'edges' of the model space (weather in the southern hemisphere, deeper ground and water layers).

One might now imagine that the development of a finer lattice, a still better model, or an increase in the computational capacity of the supercomputer, could lead to an improved success rate of almost 100%.

But in fact this belief in computability, that by setting up all parameters the behaviour of a complex system can be understood, is fallacious – though it is encountered throughout science. This is true of meteorology, physics, and other disciplines. The discovery of chaos theory has cast great doubt upon the scientific principle of the 'computability of the world'. Let us look again at the precise situation on the 23/24 July, to find out what the weather has to do with the concept of chaos.

On 23 July at 11.30 the weather report for 24 July 1985 was dictated by the duty meteorologist on to the teleprinter. According to this, the next day would be sunny and warm. As a rule, the punched tape with the weather report remains untransmitted in the teleprinter, so that later weather changes can be incorporated. The colleague who took over around midday faced a new situation, which led to second thoughts.

A drop in pressure had suddenly appeared to the west of Ireland. Such a tendency often leads to the development of a trough, leading to a worsening of the weather in the direction that the air is moving. In this case the trough was known to be capable of development, and its associated warm/cold front would then move east. This would lead to an air flow in the direction of the North Sea coast over Jutland, passing over the Baltic Sea. The duty meteorologist changed the previously prepared report and put the following message on the teleprinter:

```
23.7.1985
to newspapers
headline: correction
```

```
weather conditions.
westerly winds are carrying massive warm and cloudy air-masses
with associated disturbances towards north-west germany.
```

```
weather forecast for the weser-ems region wednesday:
heavy cloud, and intermittent rain or showers especially near
the coast.  clouds dispersing in the afternoon leading to
cessation of preciptation.  Warming to around 19 degrees
celsius.  cooling at night to 13 to 9 degrees celsius.
moderate to strong westerly winds.
further outlook:
thursday cloudy, becoming warm, and mainly dry.  temperatures
increasing.
```

```
bremen weather bureau.
```

Overnight (24.7.1985, 2.00 MEST[1]), the trough became quite pronounced and lay close to the east of Scotland. According to forecast, it moved further to the east and 12 hours later was off Jutland. The cloud cover associated with the trough then stretched only into the coastal regions and brought a few drops of rain to isolated parts. Intensive and high precipitation passed over Schleswig-Holstein. In contrast, in Bremen almost the entire day was sunny. The weather reports of 24 July then showed how wrong the forecast of 23 July 1985 was.

Weather Report 24 July 1985

	8.00	sun	rain
Bremen	clear	10 h	---
Heligoland	overcast	3 h	0.1 mm (L/m^2)
Schleswig	cloudy	1/2 h	6 mm (L/m^2)

On that day the duty meteorologist[2] could soon recite her explanation of the false prediction by heart:

---

[1] Middle European Summer Time.

[2] The information was provided by the Bremen weather office on the instructions of qualified meteorologists Sabine Nasdalack and Manfred Klöppel.

'Around 12 midday we prepare the weather forecast for the next day. For Wednesday 24
July we were predicting a sunny day, which it was. Then suddenly a trough appeared on
the map and a colleague on a later shift quickly amended the forecast. It was our
misfortune that the trough of low pressure with cloud and rain passed to the North of
Bremen at a distance of about 100 km. On the North Sea coast it did rain and it was
also fairly cool. We did our best to within 100 km.'

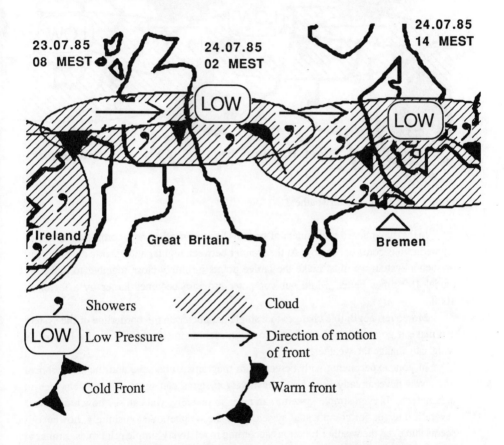

Figure 9-1 Complex weather boundaries around Bremen: 23/24.7.1985.

The weather forecast for 25 July 1985 was 'sunny and warm'. This was doubly
unfortunate for the Bremen weathermen, because on that day it then rained. This was
another case where a 'weather frontier' 50 or 100 km wide rendered all forecasts useless.
Success and failure were only a hair's breadth apart. No wonder. And the Bremen
weathermen might have realised this, if they had witnessed a related situation, which we
have 'played around with' using computer graphics.

Perhaps the Gingerbread Man had a hand in events?

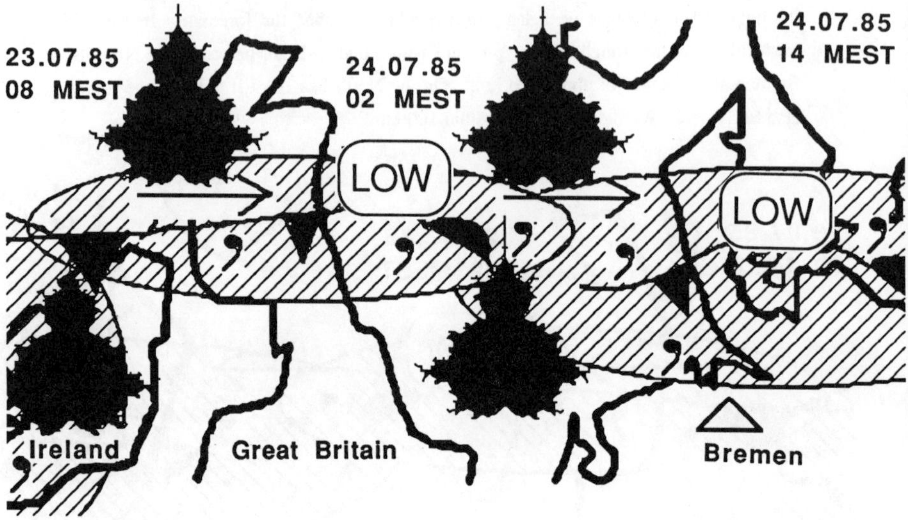

Figure 9.2 Gingerbreadweather?

In the dynamical development of systems like the weather, near complex boundaries all predictions come to nothing. At the frontier between regular and chaotic development of such a system we must make the lattice points infinitely close together to make the model 100 times better, while our computer program becomes larger by a factor of 1000...

Moreover, negligible changes to a single parameter on the borderline of the system can make it go chaotic.   In extreme form, in such a situation the 'flap of a butterfly's wing' can change the weather!

In your experiments with Feigenbaum diagrams, Julia sets, and the Gingerbread Man, you have already seen how sensitively systems can react to small changes in parameters.   If even simple systems can react so severely, you can see how much more severe it must be for complicated systems like the weather. Astonishingly, however, it seems that even the weather functions according to relatively simple principles, similar to those we have studied. This also applies to other systems whose behaviour depends on many parameters. This explains the interest of many scientists in chaos theory. While there can be no return to the ancient dream of 'computability of the world', at least there is progress in understanding how the change from an ordered computable state into an unpredictable chaotic state can occur.

In particular, chaos theory shows that there is a fundamental limit to the 'computability of the world'. Even with the best supercomputers and rapid technological development, in some cases it may be impossible to predict the behaviour of a system. It might be a political system, economic, physical or otherwise.   If we find ourselves in these boundary regions, then any further expenditure is a waste of money.

We nevertheless have a chance. We must learn to understand the 'transition from order to chaos', and to do this we begin with simple systems. At the moment this is done mostly by mathematicians and physicists. You yourself have done it too in your computer graphics experiments – of course without studying intensively the regularities towards which such a process leads.

The final aim of all this investigation is to obtain something like a 'fingerprint' of chaos. Perhaps there is a central mathematical relationship, hidden in a natural constant or a figure like the Gingerbread Man? Perhaps there are 'seismic events' that proclaim chaos? That is what we are seeking. In fact there do seem to be such signs of chaotic phenomena, which signal the transition from order to chaos.

Each of us has a rudimentary understanding of this opposing pair of concepts. We speak of order when 'everything is in its place'. Chaos, according to the dictionary, means among other things 'confusion'. We find another interesting hint there too. By 'chaos', the ancient Greeks meant the primordial substance out of which the world is built. More and more, scientists are coming to the conclusion that chaos is the normal course of events. The much-prized and well-understood order of things is just a special case. This exceptional circumstance has been the centre of scientific interest for centuries, because it is easier to understand and to use. In combination with the indisputable successes of modern science over the past 200 years it has led to a disastrous misconception: that everything is computable. When today a model fails and predicted events do not occur, we simply assume that the model is not good enough, and that that is why the predictions fail. We confidently believe that this can be corrected by more and better measurements and bigger mathematical and computational investment. At first it seems quite a startling idea that there exist problems that simply are not computable, because they lead to 'mathematical chaos'. We have discussed the example of weather at the beginning of this book and in this chapter. When two cars pass each other in the street, a vortex forms in the air between them. As a result, depending on whether we drive on the left or right, whether we live in the northern or southern hemispheres, the global high and low pressure systems are strengthened or weakened. But which weather forecasts take traffic into account? In an extreme case we speak of the 'butterfly effect'. The flap of a butterfly's wing can change our weather! This idea occurred to the American meteorologist Edward N. Lorenz in 1963. He came to the conclusion that long-term weather prediction may not be possible (compare §3.3).

In daily life and science we assume that nothing happens without a reason (the causality principle).   Whether we apply this principle to reality or to a mathematical experiment, it makes no difference. This long-standing principle that 'equal causes have equal effects', lies at the heart of our scientific thinking. No experiment would be reproducible if it did not hold. And therein lies the problem. This principle makes no statement about how small changes in causes change their effects. Even the flap of a butterfly's wing can lead to a different outcome. Usually it 'still works', and we obtain similar conclusions from similar assumptions. But often this strengthening of the causality principle fails and the consequences differ considerably. Then we speak of 'chaos'. For example, it shocks a physicist to learn that in fact the strong causality principle need not hold without restriction in classical mechanics: we need only think of a billiard player. The movement of the ball is governed by the classical laws of motion and the reflection principle. But the path of the ball is in no way predictable. After just a few bounces it can no longer be foreseen. Even a manifestly deterministic system can be chaotic. Tiny changes in the forces during the motion, or in initial conditions, lead to unpredictable movements. This can even happen as a result of the gravitational attraction of a spectator standing near the billiard table.

Chaotic systems are the rule in our world, not the exception! This opens up a reversal of our thinking:

- Chaotic systems lie behind significant dependence on initial conditions.
- The strong causality principle does not always hold. Similar causes do not always produce similar effects.
- The long-term behaviour of a system may be uncomputable.

For further examples of chaotic systems we do not need to restrict ourselves to the weather. Simpler systems can also demonstrate the limits to computability.

Time is fundamental to our civilisation. For a long time its measurement depended on the regular swing of a pendulum. Every young physics student learns the ideas that govern the pendulum. Who would imagine that chaos sets in if we add a second pendulum to its end? But that's what happens! Two double pendulums, constructed to be as similar as possible, and started with the same initial conditions, at first display similar behaviour. But quite soon we reach a situation where one of the pendulums is in unstable equilibrium and must decide: do I fall to the left or right? In this situation the system is so sensititve that even the attractive force of a passing bird, the noise of an exhaust, or the cough of an experimentalist can cause the two systems to follow entirely different paths. Many other examples present themselves if we consider flows. The air currents behind buildings or vehicles are chaotic. The eddies in flowing water cannot be pre-computed. Even in the dripping of a tap we find all behaviour from 'purest order' to 'purest chaos'. But there is a further discovery, first made in recent times: whenever a system exhibits both order and chaos, the transition occurs in the same simple manner, the one we were led to in 'step by step into chaos'. Perhaps this simple pattern is one of the first fingerprints of chaos, the first 'seismic event', which signals uncomputability.

What do water drops, heartbeats, and the arms race, have in common? 'Nothing', anyone would probably say, if asked. But all three things are involved in chaos research; all three show the same simple pattern that we saw in 'step by step into chaos'.

The Dripping Tap

Chaos can be found in a dripping tap. Here we can even experiment for ourselves. Normally a tap has two states: open or closed. We are of course interested in the border zone, when the tap is only partially open. The dripping tap then represents an ordered system. The drops are of equal size and fall at regular intervals. If the tap is opened a little more, the drops fall more quickly, until we encounter a phenomenon which we have previously seen only as mathematical feedback: suddenly two drops of different sizes appear, one after the other.

In the same way that a curve in the Feigenbaum diagram can grow two branches, where the sequence alternates between a small and a large value, so also does the water behave. After a large drop there follows a smaller one; after a small one comes a large one.

Unfortunately it is not so easy to observe what happens next. The very rapid events are best seen using photography or with the aid of a stroboscope. Then under some conditions we can see a further period–doubling: a regular sequence of four different drops! Of course, the periods of time between them also differ.

With an accurate timer all this can be made quantitative. The arrangement should be one where the flow rate of the water can be changed reproducibly. Unfortunately a tap is not accurate enough, and it is better to use a water reservoir about the size of an aquarium. A tube is placed across the edge and ends in a spout, made as uniform as possible. It should point downwards. The height of this spout, compared to the water level, controls the quantity of fluid that flows per minute. Because the size of drops is hard to measure, we measure the times at which the drops fall. To do this we let them fall through a suitable light–beam, a few centimetres below the opening. An electronic timer produces a sequence of measurements. We discover:

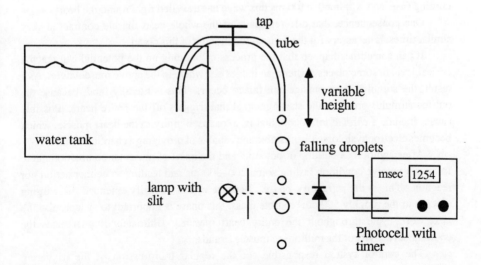

Figure 9–3   Water drop experiment.

- If a little water flows, the drops follow each other at equal intervals.
- By increasing the flow rate, periods 2 and 4 can be clearly detected.
- Finally, shortly before the water flows in a smooth stream, big and small drops follow each other indiscriminately. We have reached chaos!

A possible design of apparatus is shown in Figure 9-3.

The Heartbeat

The heartbeat is another process, in which the complex system 'heart' attains a chaotic state. We consider the process somewhat more precisely.

Normal heartbeats depend upon the synchronised behaviour of many millions of strands of heart muscle. Each of these individual elements of the heart's musculature runs through an electrophysiological cycle lasting about 750 ms. Sodium, potassium and chlorine ions are so arranged, inside and outside the cell wall, that by a combination of chemical events each individiual cell attains a progressively unstable 'explosive' electrophysiological state. Millions of tiny chemical 'catapults' are sprung.

It is now purely a matter of time before this situation leads to a sudden increase in the electrical potential difference across the cell membrane. This is the trigger for the muscular activity of the heart. Normally the initiative for this is controlled by natural pacemaker tissue, which conveys the impulse to the active muscles by a circuit similar to the nervous system.

Once the above chain reaction has begun, it propagates through the musculature of the heart by transmission from each cell to its neighbours. The 'explosive' neighbouring cells are triggered. The process of transmission takes the form of a travelling wave, like a burning fuse, and within 60 - 100 ms this wave has travelled right round the heart.

One consequence, that different parts of the whole heart muscle contract at very similar times, is necessary for the optimal functioning of this organ.

It can sometimes happen that the process of building up the potential can reach a critical level in some places earlier than it does elsewhere in the active musculature. As a result, the stimulus from outside can fail to occur. This is literally fatal, because the outside stimulus provides the starting signal that triggers off the entire heart. Possible causes include a *cardiac infarction*, that is, a localised injury to the heart muscle, which becomes electrophysiologically unstable and capable of providing a chaotic stimulus.

After 'ignition', a buildup of potential and the onset of muscular contraction, there follows a passive condition lasting some 200–300 ms and leading to neither action nor reaction. This is the *refractory phase*. Because of this spatially extended discharging process in the roughly 500 cm^3 muscle mass, this phase is important to synchronise the next work–cycle throughout the entire heart muscle. Ultimately it guarantees the coordinated activity of the millions of muscle strands.

The control system responsible for the regular transmission of the triggering impulses is essential. If a 'firing' impulse arrives at the active muscles too soon, then these will react too soon. The result is an irregular heartbeat: an *extrasystole*.

This is primarily a normal phenomenon. Even in a healthy heart it can be observed, insofar as this impulse may act on a uniformly ready muscle. But if parts of the heart find themselves in an inaccessible stationary state, while other parts are ready to transmit or propagate a spreading impulse, then a fatal condition ensues. It begins with a division of the activity of the heart muscle into regions that are no longer in synchrony. A consequence is the threat that islands of musculature will exhibit cyclic stimulus processes: chaos breaks out, and the result is *fibrillation*. Despite maximal energy use, no output occurs from the biological pump 'heart'. A vicious circle has set in.

Figure 9-4 shows a normal ECG and an ECG showing the typical symptoms of fibrillation. The amplitude is exaggerated compared with the time in relative units.

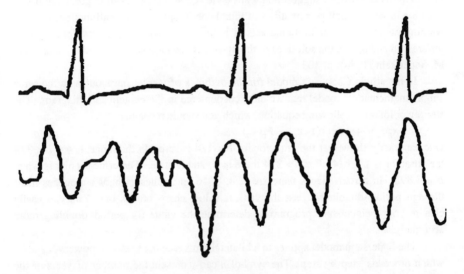

Figure 9-4   ECG curves: normal action (top) and fibrillation.

Astonishingly, this transition from order to chaos, which means fibrillation and death, seems to follow the same simple pattern as water drops and the Feigenbaum diagram. Feigenbaum called the phase shortly before chaos set in *period-doubling*. In this situation complex systems oscillate to and fro between 2, then 4, then 8, then 16, and so on states. Finally nothing regular can be observed.

In the case of the heart, Richard J. Cohen of the Massachusetts Institute of Technology found the same typical pattern in animal experiments. These might imply that an early warning of such extrasystoles in the heart might perhaps be able to save lives.

Outbreaks of War
As a final example we mention that researchers in peace studies have also shown an interest in chaos theory. Again we can see the first tentative beginnings, which as in other examples suggest how effective the applications might become. But it is

questionable whether so simplified a formulation of this fundamental problem is really applicable to our times.

Today we live in a world of the balance of terror. Complex systems such as the global environment or national defence switch between phases of order and disorder. Chaotic situations are well known in both systems. Seveso, Bhopal and Chernobyl are environmental catastrophes. World wars and other more local conflicts are peace catastrophes.

History offers many examples of how the precursors of an outbreak of war announce themselves. At the threshold between war and peace political control and predictability are lost. 'I suggest that war be viewed as a breakdown in predictability: a situation in which small perturbations of initial conditions, such as malfunctions of early warning radar systems or irrational acts of individuals disobeying orders, lead to large unforeseen changes in the solutions to the dynamical equations of the model,' says Alvin M. Saperstein (1984), p. 303.

In his article *Chaos - a model for the outbreak of war* he starts out with the same simple mathematical model that we have encountered in the Feigenbaum diagram. The starting point is the following equation, which you should recognise:

$$x_{n+1} = 4*b*x_n*(1-x_n) = F_b(x_n).$$

One can easily show that the attractor for $b < 1/4$ is zero. In the range $1/4 < b < 3/4$ the attractor is $1 - b/4$. For $b > 3/4$ there is no stable state. The critical point is when $b = 0.892$. In certain places there are 2, 4, 8, 16, etc. states. Chaos announces itself through period-doubling. Past the critical value, chaos breaks out. You can easily change your Feigenbaum program to determine the value for period-doubling more accurately.

The Saperstein model applies to a bilateral arms race between two powers X and Y, which proceeds 'step by step'. The symbol n can represent the number of years or the military budget. The model of the arms race consists of the following system:

$$x_{n+1} = 4*a*y_n*(1-y_n) = F_a(y_n)$$
$$y_{n+1} = 4*b*x_n*(1-x_n) = F_b(x_n)$$

with $0 < a,b < 1$. The dependent variables x_n and y_n represent the proportion that the two nations spend on their armaments. Thus the following should hold: $x_n > 0$ and $y_n < 1$.

The armament behaviour of one nation depends of course on that of its opponent up to the current time, and conversely. This is expressed by the dependence between x and y. Much as in the measles example, the factors $(1-y_n)$ and $(1-x_n)$ give the proportion of gross national product that is not invested in armaments. Depending on the values of the parameters a and b, we get stable or unstable behaviour of the system. The two nations can 'calculate' the behaviour of their opponents - or perhaps not.

Saperstein has derived a table from the book *European Historical Statistics 1750-1970*, which shows the military output of several countries between 1934 and 1937 in terms of gross national product (GNP).

	France	Germany	Italy	UK	USSR
1934	0.0276	0.0104	0.0443	0.0202	0.0501
1935	0.0293	0.0125	0.0461	0.0240	0.0552
1936	0.0194	0.0298	0.0296	0.0781	not known
1937	0.0248	0.0359	0.0454	0.0947	not known

Table 9-1 Dependence of military output on GNP, Saperstein (1984) p. 305.

The values in Table 9-1 are now taken in turn for x_0,y_0, respectively x_1,y_1, in the model equations, and a and b determined from them. Table 9-2 shows the result:

		a	b
France–Germany	(1934–35)	0.712	0.116
France–Italy	(1936–37)	0.214	0.472
UK–Germany	(1934–35)	0.582	0.158
UK–Italy	(1934–35)	0.142	0.582
USSR–Germany	(1934–35)	1.34	0.0657
USSR–Italy	(1936–37)	0.819	0.125

Table 9-2 Parameters a and b, Saperstein (1984), p. 305.

Accordingly, the arms race between USSR and Germany during the chosen period is in the chaotic region. France–Germany and USSR–Italy are, in Saperstein's interpretation, near the critical point.

For such a simple model the results are rather surprising, even if no temporal or historical development is taken into consideration. We suggest that you test the Saperstein model and locate new data. Statistical yearbooks can be found in most libraries. Of course the model is too simple to provide reliable results. We are in the early days of chaos research, and we can only hope that eventually it will be possible to help avoid chaotic situations in real life.

Phase Transitions and Gipsy Moths

These three examples are typical of many other phenomena currently under investigation by scientists in many fields. We briefly describe one of them.

Changes from order to disorder are on the physicist's daily agenda. The transition from water to steam, the change from a conducting to a superconducting state at low temperature, the transition from laminar to turbulent flow or from solid to fluid or gaseous states characterise such *phase transitions*. Phase transitions are extremely complex, because in the system certain elements 'wander' for indeterminate times, oscillating to and fro between two possible states of the system. Mixtures of states can develop, without the system as a whole becoming chaotic.

Of great interest today is the manufacture of new materials. Their magnetic and non-magnetic properties play an important role for different characteristics, such as elasticity or other physical properties. In fact, today, chaos theorists are already 'experimenting' with such materials, albeit hypothetical ones. For this purpose the data of real materials are incorporated in a computer model, which calculates the magnetic and non-magnetic zones. The pictures obtained in this manner are similar to Julia sets. Of course people are particularly interested in the complex boundaries between the two regions.

A surprising result of this research could be the discovery that highly complex phase transitions, as they apply to magnetism, can be understood through simple mechanisms.

Another example from biology shows that even here the Feigenbaum scenario is involved. The biologist Robert M. May investigated the growth of a type of insect, the gipsy moth *Lymantria dispar*, which infests large areas of woodland in the USA. In fact there are all the signs that this sometimes chaotic insect behaviour can be described by the Feigenbaum formula: see May (1976), Breuer (1985).

Chaos theory today concerns itself, as we have seen, with the broad question of how the transition from order to chaos takes place. In particular there are four questions that the researchers pose:

- How can we detect the way in which 'step by step' inevitably leads to chaos?

    Is there a fingerprint of chaos, a characteristic pattern or symptom, which can give advance warning?

- Can this process be formulated and pinned down in simple mathematical terms?

    Are there basic forms, such as the Gingerbread Man, which always occur in different complex systems of an economic, political, or scientific nature?

    Are there basic relations in the form of system constraints or invariants?

- What implications do all these discoveries have for the traditional scientific paradigm?

    What modifications or extensions of existing theoretical constructions are useful or important?

- How do natural systems behave in transition from order to chaos?

Video Feedback

To bring to a close this stepwise excursion into the continuing and unknown development of chaos theory, we will introduce you to a graphical experiment, which will set up the beginnings of a visual journey into the 'Land of Infinite Structures'. This experiment requires apparatus which is already available in many schools and private households. The materials are: a television set, a video camera, and a tripod. Whether you use a black-and-white or a colour TV is unimportant.

The setup is conceptually simple. The video camera is connected to the TV and pointed at the screen. Both machines are switched on. The camera films the picture that it itself creates. In the TV we see a picture of the TV, containing a picture of the TV, containing a picture of the TV, containing...

A recursive, self–similar picture appears. Now bring the camera nearer to the screen. At certain places the screen becomes very bright, at others it stays dark. Feedback leads to a strengthening, in one or the other direction. The interesting thing for us is the border between the light and dark regions. There accidental variations are enhanced, so that scintillations appear, the first symptoms of chaos.

A word about the parameters we can vary in this experiment. On the TV these are brightness and contrast; on the camera they are field of vision (zoom) and sensitivity. If necessary the automatic exposure control must be shut off. Our third piece of apparatus plays an especially important role: the tripod. With this we can tilt the camera at an angle between 0° and 90° above the monitor. A bright speck on the screen can become brighter after being filmed and may strengthen further iterations.

Switch off the light and darken the room. Suddenly the journey begins. Through feedback between the camera and the TV, remarkable dynamically changing structures appear (Figure 9–5), and chaos shows its face. The patterns that arise in this way are so varied that we will make no attempt to describe them.

Nobody can yet compute or predict these structures. Why should it be otherwise?

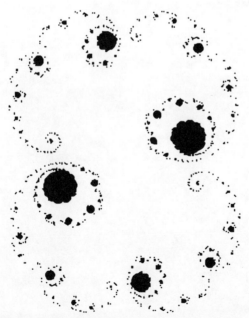

Figure 9–5 Revolving, self–modifying patterns on the screen.[3]

---

[3]The pictures on the TV screen generally resemble such computer–generated structures.

10 Journey to the Land of Infinite Structures

With our final experiments on fractal graphics, and questions of the fundamental principles of chaos theory that are today unresolved, we end the computer graphics experiments in this book. That does not mean that the experiment is over for you. On the contrary, perhaps it has only just really begun.

For the Grand Finale we invite you on a trip into the Land of Infinite Structures. That is what we have been discussing all along. A microcosm within mathematics, whose self–similar structures run to infinity, has opened up before us.

Have you ever seen the Grand Canyon or Monument Valley in America? Have you perhaps flown in an aeroplane through their ravines and valleys? Have you gazed out from a high peak upon the scenery below? We have done all that in the Land of Infinite Structures, too. And here we have once more collected together some of the photographs as a memento.

With the sun low in the evening, when the contours are at their sharpest, we fly from the west into Monument Valley. Rocky outcrop upon rocky outcrop towers red into the sky. Below spreads the flat land of the reservation. Between two mesas we lose height and turn to the right:

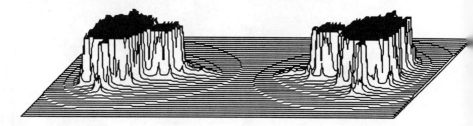

Figure   10.1

Ahead of us stretches the plateau. At its edge, as far as the eye can see in the twilight, stretch the cliffs, seeming to lose themselves in the infinite distance:

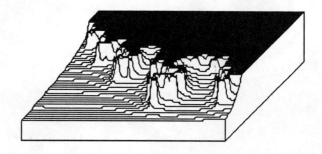

Figure   10.2

A glance at our flight-plan shows that we are flying in the area of a basin boundary:

Figure 10.3

The nearer we approach, the more precipitous and forbidding the slopes become, throwing long shadows in the evening sun:

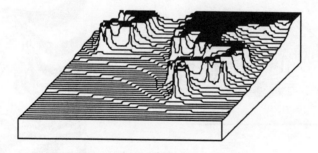

Figure 10.4

Suddenly a gap opens up in the hitherto impenetrable massif. We follow it with the setting sun. The gap becomes smaller and smaller. We switch on the terrain radar. Whenever critical points loom ahead, we switch to the next level of magnification:

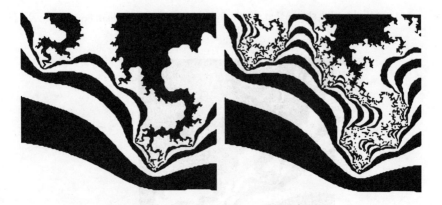

Figure   10.5

Once more we must climb, to gain height, until behind the winding mountain chain the
airport lights appear:

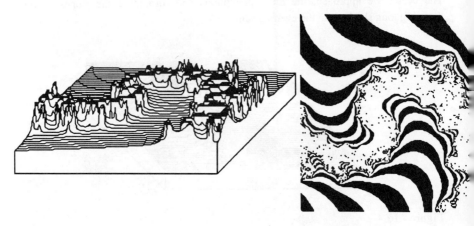

Figure   10.6   Figure 10.7

The following day, when we fly over the Grand Canyon, our appetite for the variety of
the forms that the Colorado has engraved in the rock during its thousand-year efforts is
insatiable. Joe, the pilot, has switched on the terrain radar, which plots contour lines.
Remarkably, as in the three-dimensional world of forms, which lies before our eyes, a
strange reality now appears.

Figure   10.8

Figure   10.9   Figure 10.10

Figure   10.11   Figure 10.12

Figure   10.13

Figure   10.14

Figure   10.15

Figure   10.16

A few days later we depart from base in a car to see the local terrain. We encounter two Indians, who want to sell us their artwork, carvings, drawings, and finely beaten jewellery. We have never seen such art before. The forms are hardly ever geometrically simple. Indeed, in general we find complex patterns, which split up infinitely finely (see the following figures). We ask the Indians what this art-form is called but do not understand what they say. 'Ailuj, Ailuj!' they cry, and point to their drawings. What is this 'Ailuj'? We do not understand.

Figure   10.17

Figure   10.18

On the next day of the trip we once more fly over the outcroppings of the Grand Canyon.
'How many years did it take the Colorado to produce these contours? It is at most a
hundred years since humans first saw them.'

We scramble out of the aircraft and look around at the place we have reached:

Figure   10.19

Figure   10.20

Figure   10.21

We look again. I point silently to the heavens. A flock of birds flies to the west. They go on forever.

11 Building Blocks for Graphical Experiments

11.1   The Fundamental Algorithms

In the first eight chapters we have explained to you some interesting problems and formulated a large number of exercises, to stimulate your taste for computer graphics experiments. In the chapters following Chapter 11 we provide solutions to the exercises, sometimes as complete programs or else as fragments. The solution, the complete Pascal program, can be obtained by combining the ready-made component fragments, after which you can perform a great number of graphical experiments.

Structured programs are programs in which the founder of the programming language Pascal – the Swiss computer scientist Niklaus Wirth – has always taken a great interest. Such programs are composed only of procedures and functions. The procedures are at most a page long. Naturally such programs are well documented. Any user who reads these programs can understand what they do. The variable names, for example, convey their meaning and are commented at length when necessary. In particular the programs are written in a structured fashion, in which the 'indentation rules' and 'style rules' for Pascal are strictly adhered to. We hope that you too will write structured programs, and we would like to offer some advice.

Perhaps you have read the entire book systematically up to Chapter 11, to get a good survey of the formulation of the problem; perhaps you have also written the first few programs. By analysing our program description you have surely noticed that the structure of our programs is always the same. The reason is that we have always tried to write structured programs.

These are, in particular:

- portable, so that they can easily run on other computers ('machine-independent');
- clearly structured ('small procedures');
- well commented ('use of comments');
- devoid of arbitrary names ('description of variables and procedures').

Exceptions are limited and removed in a tolerable time.

New programs can quickly be assembled from the building blocks. You can get a clear idea of the structure of our programs from Example 11.1-1:

Reference Program 11.1-1 (cf. Program 2.1-1)

```
PROGRAM EmptyApplicationShell; (* for computer XYZ        *)
    (* library declarations where applicable              *)

    CONST
        Xscreen = 320;    (* e.g. 320 points in x-direction *)
        Yscreen = 200;    (* e.g. 200 points in y-direction *)

    VAR
        PictureName : string;
```

```
      Left, Right, Top, Bottom : real;
      (* include additional global variables here *)
      Feedback : real;
      Visible, Invisible : integer;

(* ----------------------UTILITY-------------------------*)
(* BEGIN:  Useful Subroutines *)
   PROCEDURE ReadReal (information : STRING; VAR value
                       : real);
   BEGIN
      write (information);
      readln (value);
   END;


   PROCEDURE ReadInteger (information : STRING; VAR value
                       : integer);
   BEGIN
      write (information);
      readln (value);
   END;


   PROCEDURE ReadString (information : STRING; VAR value
                       : string);
   BEGIN
      write (information);
      readln (value);
   END;


   PROCEDURE InfoOutput (information : STRING);
   BEGIN
      writeln (information);
      writeln;
   END;


   PROCEDURE CarryOn (information : STRING);
   BEGIN
      write (information, ' hit <RETURN>');
      readln
   END;


   PROCEDURE CarryOnIfKey;
```

```
    BEGIN
      REPEAT
      UNTIL KeyPressed (* DANGER - machine-dependent!          *)
    END;

    PROCEDURE Newlines (n : integer);
      VAR
         i : integer;
    BEGIN
      FOR i := 1 TO n DO
         writeln;
    END;

(* END: Useful Subroutines *)
(* ------------------------UTILITY------------------------*)

(* ------------------------GRAPHICS------------------------*)
(* BEGIN:  Graphics Procedures *)

    PROCEDURE SetPoint (xs, ys : integer);
    BEGIN
      (*  Insert machine-specific graphics commands here     *)
    END;

    PROCEDURE SetUniversalPoint (xw, yw: real);
      VAR xs, ys : real;
    BEGIN
      xs := (xw - Left) * Xscreen / (Right - Left);
      ys := (yw - Bottom) * Yscreen / (Top - Bottom);
      SetPoint (round(xs), round(ys));
    END;

    PROCEDURE GoToPoint (xs, ys : integer);
    BEGIN
      (* move without drawing *)
      (*  Insert machine-specific graphics commands here     *)
    END;

    PROCEDURE DrawLine (xs, ys : integer);
    BEGIN
      (*  Insert machine-specific graphics commands here     *)
    END;
```

```
PROCEDURE DrawUniversalLine (xw, yw : real);
   VAR xs, ys : real;
BEGIN
   xs := (xw - Left) * Xscreen/(Right - Left);
   ys := (yw - Bottom) * Yscreen / (Top - Bottom);
   DrawLine (round(xs), round(ys));
END;


PROCEDURE TextMode;
BEGIN
   (* switch on text-representation *)
   (*  Insert machine-specific graphics commands here    *)
END;


PROCEDURE GraphicsMode;
BEGIN
   (* switch on graphics-representation *)
   (*  Insert machine-specific graphics commands here    *)
END;


PROCEDURE EnterGraphics;
BEGIN
   writeln ('To end drawing hit <RETURN> ');
   write ('now hit <RETURN> '); readln;
   GraphicsMode;
END;


PROCEDURE ExitGraphics
BEGIN
(* machine-specific actions to exit from Graphics Mode    *)
   TextMode;
END;

(END: Graphics Procedures *)
(------------------------GRAPHICS------------------------*)

(---------------------APPLICATION------------------------*)
(BEGIN: Problem-specific procedures *)
(* useful functions for the given application problem         *)

   FUNCTION f(p, k : real) : real;
```

```
BEGIN f := p + k * p * (1-p);
END;


PROCEDURE FeigenbaumIteration;
   VAR
      range, i: integer;
      population, deltaxPerPixel : real;
   BEGIN
      deltaxPerPixel := (Right - Left) / Xscreen;
      FOR range := 0 TO Xscreen DO
         BEGIN
            Feedback := Left + range * deltaxPerPixel;
            population := 0.3;
            FOR i := 0 TO invisible DO
               population := f(population, Feedback);
            FOR i := 0 TO visible DO
               BEGIN
                  SetUniversalPoint (Feedback, population);
                  population := f(population, feedback);
               END;
         END;
   END;


(* END: Problem-specific procedures                              *)
(*-----------------------APPLICATION----------------------*)
(*-----------------------MAIN----------------------------*)
(* BEGIN: Procedures of Main Program                            *)

   PROCEDURE Hello;
   BEGIN
      TextMode;
      InfoOutput ('Calculation of                       ');
      InfoOutput ('-------------------------');
      Newlines (2);
      CarryOn ('Start :');
      Newlines (2);
   END;


   PROCEDURE Goodbye;
   BEGIN
      CarryOn ('To stop : ');
   END;
```

```
PROCEDURE Initialise;
    BEGIN
        ReadReal ('Left                            >', Left);
        ReadReal ('Right                           >', Right);
        ReadReal ('Top                             >', Top);
        ReadReal ('Bottom                          >', Bottom);
        ReadInteger ('Invisible                    >', invisible);
        ReadInteger ('Visible                       >', visible);
            (* possibly further inputs *)
        ReadString ('Name of Picture               >',PictureName);
    END;


    PROCEDURE ComputeAndDisplay;
    BEGIN
        EnterGraphics;
        FeigenbaumIteration;
        ExitGraphics;
    END;

(* END: Procedures of Main Program                                  *)
(* ---------------------------MAIN------------------------- *)
    BEGIN (* Main Program *)
        Hello;
        Initialise;
        ComputeAndDisplay;
        Goodbye;
    END.
```

1 Structure of Pascal Programs

All program examples are constructed according to the following scheme:

```
PROGRAM ProgramName;
    |   Here follows the declaration part including
    |   • Library declarations (if applicable)
    |   • Constant declarations
    |   • Type declarations
    |   • Declaration of global variables
(*---------------------------------------------------------- *)
(* BEGIN: Useful Subroutines                                        *)
    |   • Here follows the definition of the useful subroutines
(* END: Useful Subroutines                                          *)
```

```
(*-------------------------------------------------------------- *)
(* BEGIN: Graphics Procedures                                    *)
    |  • Here follows the definition of the graphics pProcedures
(* END: Graphics Procedures                                      *)
(*-------------------------------------------------------------- *)
(* BEGIN: Problem-specific Procedures                            *)
    |  • Here follows the definition of the problem-specific procedures
(* END: Problem-specific Procedures                             *)
(*-------------------------------------------------------------- *)
(* BEGIN: Procedures of Main Program                             *)
    |  • Here follows the definition of the
    |  • Procedures of the main program:
    |  • Hello, Goodbye, Initialise, ComputeAndDisplay
(* END: Procedures of Main Program                              *)
(*-------------------------------------------------------------- *)
    BEGIN (* Main Program *)
       Hello;
       Initialise;
       ComputeAndDisplay;
       Goodbye;
    END.
```

2 Layout of Pascal Programs

All Pascal programs have a unified appearance.

- Global symbols begin with a capital letter. These are the name of the main program, global variables, and global procedures.

- Local symbols begin with a lower-case letter. These are names of local procedures and local variables.

- Keywords in Pascal are written in capitals or printed boldface.

3 Machine-Independence of Pascal Programs

By observing a few simple rules, all Pascal programs can be used on different machines. Of course today's computers unfortunately still differ widely from one another. For this reason we have set up the basic structure of our reference program (see overleaf), so that the machine-dependent parts can quickly be transported to other machines. Model programs and reference programs for different makes of machine and programming languages can be found in Chapter 12.

We now describe in more detail the overall structure of Pascal programs.

Global Variables

The definitions of global constants, types, and variables are identified with bold capital headings for each new global quantity. Although not all of them are used in the same program, a typical declaration for a Julia or a Gingerbread Man program might look like this:

Program Fragment 11.1-2

```
CONST
    Xscreen = 320;              (*screen width in pixels     *)
    Yscreen = 200;              (*screen height in pixels    *)
    WindowBorder = 20;          (*0 for a Macintosh          *)
    Limit = 100;                (*test for end of iteration  *)
    Pi = 3.141592653589;        (*implemented in many dialects*)
    Parts = 64;                 (* for fractal landscapes    *)
    PartsPlus2 = 66:            (*=Parts + 2, for landscapes *)


TYPE
    IntFile = FILE OF integer;  (*machine-independent        *)
    CharFile = Text;            (*storage of picture data    *)


VAR
    Visible, Invisible,         (*drawing limits for         *)
                                (*Feigenbaum diagrams        *)
    MaximalIteration, Bound,    (*drawing limits for         *)
                                (*Julia and Mandelbrot sets  *)
    Turtleangle, Turtlex, Turtley,
    Startx, Starty, Direction,
    Depth, Side,                (*turtle graphics            *)
    Xcentre, Ycentre, Radius,   (*screen parameters of       *)
                                (*Riemann sphere             *)
    Quantity, Colour, Loaded,   (*Data values for screen-    *)
                                (*independent picture storage *)
    Initial, Factor,            (*fractal mountains          *)
    D3factor, D3xstep, D3ystep,     (*3D-specialities        *)
                        :integer;
Ch                                      :char;
PictureName, FileName   :STRING;
    Left, Right, Top, Bottom,   (*screen window limits       *)
    Feedback, Population,       (*parameters for Feigenbaum  *)
    N1, N2, N3, StartValue      (*parameters for Newton      *)
    Creal, Cimaginary,          (*components of c            *)
```

```
FixValue1, FixValue2        (*tomogram parameters         *)
Width, Length,              (*Riemann sphere coordinates  *)
Power,                      (*for generalised Gingerbread *)
HalfPi,                     (* used for arc sine, = Pi/2   *)
                        :real;
F, In, Out              : IntFile;
InText, OutText         : CharFile;
                            (* for screen-independent      *)
                            (* data storage                *)
Value                   : ARRAY [0..Parts, 0..PartsPlus2] OF
                            integer;
                            (* fractal landscapes          *)
CharTable               :ARRAY [0..63] OF Char;
                            (* look-up tables              *)
IntTable                : ARRAY ['0'..'z'] OF integer;
D3max                   : D3maxType;
                            (* maximal values for 3D        *)
```

Graphics   Procedures

With graphics procedures the problem of machine–dependence becomes prominent. The main procedures for transforming from universal to picture coordinates are self–explanatory.

Problem–specific   Procedures

Despite the name, there are no problems with problem–specific procedures. On any given occasion the appropriate procedures and functions are inserted in the declaration when called from the main program.

Useful   Subroutines

In example program 11.1–1 we first formulate all the useful subroutines required for the program fragments of previous chapters; but these are not described in detail. This would probably be superfluous, since anyone with a basic knowldedge of Pascal or other programming languages can see immediately what they do.

Most of these procedures rely upon reading data from the keyboard, whose input is accompanied by the output of some prompt. In this way any user who has not written the program knows which input is required. The input of data can still lead to problems if this simple procedure is not carried out in the manner that the programmer intended. If the user types in a letter instead of a digit, in many dialects Pascal stops with an unfriendly error message. We recommend that you add to the basic algorithms procedures to protect against input errors. You can find examples in many books on Pascal.

We can now finish the survey of the structure and basic algorithms of our reference program, and in the next chapter we lay out solutions to problems that were not fully explained earlier.

Since the program structure is always the same, to provide solutions we need only give the following parts of the program:

- The problem–specific part.
- The input procedure.

From the input procedure you can read off without difficulty which global variables must be declared.

11.2   Fractals   Revisited

Now we begin the discussion of the solutions to the exercises, supplementing the individual chapters.   Occasionally we make a more systematic study of things that are explained in the first eight chapters. Why not begin at the end, which is likely to be still fresh in your mind?

Do you remember the 'fractal computer graphics' from Chapter 8? Possible partial solutions for the exercises listed there are here given as program fragments. In the main program the appropriate procedures must be called in the places signified. This type of sketch is very quick to set up.

Program Fragment 11.2-1 (for Chapter 8)

```
. . .
    VAR
    (* insert further global variables here *)
        Turtleangle, Turtlex, Turtley : integer;
        Startx, Starty, Direction, Depth, Side : integer;
    . . .
(*----------------------APPLICATION----------------------*)
(* BEGIN: Problem-specific procedures *)
(*Here follow the functions needed in the application program*)

    PROCEDURE Forward (step : integer);
    VAR
        xStep, yStep : real;
    BEGIN
        xStep := step * cos (Turtleangle * Pi) / 180.0);
        yStep := step * sin (Turtleangle * Pi) / 180.0);
        Turtlex := Turtlex + trunc (xStep);
        Turtley := Turtley + trunc (yStep);
        DrawLine (Turtlex, Turtley);
    END;
```

```
PROCEDURE Backward (step : integer);
BEGIN
   Forward (- step);
END;


PROCEDURE Turn (alpha : integer);
BEGIN
   Turtleangle := (Turtleangle + alpha) MOD 360;
END;


PROCEDURE StartTurtle;
BEGIN
   Turtleangle := 90; Turtlex := Startx; Turtely := Starty;
   SetPoint (Startx, Starty);
END;


PROCEDURE dragon (Depth, Side : integer);
BEGIN
   IF Depth = 0 THEN
      Forward (Side)
   ELSE IF Depth > 0 THEN
      BEGIN
         dragon (Depth - 1, trunc (Side));
         Turn (90);
         dragon (-(Depth - 1), trunc (Side));
      END
   ELSE
      BEGIN
         dragon (-(Depth + 1), trunc (Side));
         Turn (270);
         dragon (Depth + 1, trunc (Side));
      END;
   END;


(* END: Problem-specific procedures *)
(*------------------------APPLICATION--------------------- *)


   ...
   PROCEDURE Initialise;
   BEGIN
      ReadInteger ('Startx    >', Startx);
      ReadInteger ('Starty    >', Starty);
```

```
      ReadInteger ('Direction >', Direction);
      ReadInteger ('Depth     >', Depth);
      ReadInteger ('Side      >', Side);
   END;


   PROCEDURE ComputeAndDisplay;
   BEGIN
      EnterGraphics;
      StartTurtle;
      dragon (Depth, Side);
      ExitGraphics;
   END;
```

The declaration part of the main program always stays the same.

Please take note of this solution, because it will run in this form on any computer, provided you insert your machine-specific commands in the appropriate places of the graphics part. The relevant global variables for all exercises are given in full. In the graphics part you must always insert your commands in the procedures SetPoint, DrawLine, TextMode, GraphicsMode, and ExitGraphics (see §11.1 and Chapter 12). We have included the implementation of our own turtle graphics in the problem-specific procedures. If your computer has its own turtle graphics system (UCSD systems, Turbo Pascal systems) you can easily omit our version. But do not forget to include the global variables of your turtle version. That applies also to the correct initialisation of the turtle, which we simulate with the aid of the procedure StartTurtle. Read the hints in Chapter 12.

All procedures which stay the same will here and in future be omitted to save space.

In Chapter 8, Figure 8.2-4 we showed a fractal landscape with mountains and seas. You must have wondered how to produce this graphic. Here you will find an essentially complete solution. Only the seas are absent.

Program 11.2-2 (for Figure 8.2-4)
```
   PROGRAM fractalLandscapes;
      CONST
         Parts = 64;
         PartsPlus2 = 66;  {= Parts + 2}
      VAR
         Initial, Picsize : integer;
         Value : ARRAY [0..Parts, 0..PartsPlus2] OF integer;
      Mini, Maxi, Factor, Left, Right, Top, Bottom : real;
   ...
(* Insert the global procedures (see 11.1-1) here              *)
```

```
(* BEGIN: Problem-specific procedures                          *)
      FUNCTION RandomChoice (a, b : integer) : integer;
(* Function with a side-effect on Maxi and Mini                *)
         VAR zw : integer;
      BEGIN
         zw := (a + b) DIV 2 + Random MOD Picsize - Initial;
         IF zw < Mini THEN Mini := zw;
         IF zw > Maxi THEN Maxi := zw;
         RandomChoice := zw;
      END;  (* of RandomChoice *)


      PROCEDURE Fill;
         VAR i, j : integer;


         PROCEDURE  full;
           VAR xko, yko : integer;
         BEGIN
           yko := 0;
           REPEAT
             xko := Initial;
             REPEAT
                Value [xko, yko] :=
                   RandomChoice (Value [xko - Initial, yko],
                                 Value [xko + Initial, yko]);
                Value [yko, yko] :=
                   RandomChoice (Value [yko, xko - Initial ],
                                 Value [yko, xko + Initial ]);
                Value [xko, Parts - xko - yko] :=
                   RandomChoice (
                      Value [xko-Initial,
                             Parts-xko-yko+Initial ],
                      Value [xko+Initial,
                             Parts-xko-yko-Initial ]);
                xko := xko + Picsize;
                UNTIL xko > (Parts - yko);
             yko := yko + Picsize;
           UNTIL yko >= Parts;
         END;  (* of full *)

      BEGIN (* of Fill *)
         FOR i := 0 TO Parts DO
            FOR j := 0 TO PartsPlus2 DO
```

```
        Value [i, j] := 0;
    Mini := 0; Maxi := 0;
    Picsize := Parts;
    Initial := Picsize DIV 2;
    REPEAT
        full;
        Picsize := Initial;
        Initial := Initial DIV 2;
    UNTIL Initial := Picsize;
    Value [0, Parts + 1] := Mini;
    Value [1, Parts + 1] := Maxi;
    Value [2, Parts + 1] := Picsize;
    Value [3, Parts + 1] := Initial;
END (* of Fill *)


PROCEDURE Draw;
    VAR xko, yko : integer;

    PROCEDURE slant;
        VAR xko : integer;
    BEGIN (* of slant *)
        SetUniversalPoint (yko, yko+Value[0,yko]*Factor);
        FOR xko := 0 TO Parts - yko DO
            DrawUniversalLine (xko+yko,
                    yko + Value[xko,yko]*Factor);
        FOR xko := Parts - yko TO Parts DO
            DrawUniversalLine (xko+yko,
                    yko + Value[Parts-yko,
                            Parts-xko]*Factor);
    END;  (* of slant *)


    PROCEDURE along;
        VAR xko : integer;
    BEGIN
        SetUniversalPoint (xko, Value[xko,0]*Factor);
        FOR yko := 0 TO Parts -xko DO
            DrawUniversalLine (xko+yko,
                    yko+Value[xko,yko]*Factor);
        FOR yko := Parts - xko TO Parts DO
            DrawUniversalLine (xko+yko,
                    xko+Value[Parts-yko, Parts-xko]*Factor);
```

```
            END;   (* of along *)

        BEGIN (* of Draw *)
           FOR yko := 0 TO Parts DO
               slant;
           FOR xko := 0 TO Parts DO
               along;
           END; (* of Draw *)

(* END: Problem-specific Procedures *)
        PROCEDURE Initalise;
        BEGIN
           ReadReal (' Left        >, Left);
           ReadReal (' Right       >', Right);
           ReadReal (' Bottom      >', Bottom);
           ReadReal (' Top         >', Top);
           ReadReal (' Factor      >', Factor);
           Newlines (2);
           InfoOutput ('wait 20 seconds ');
           Newlines (2);
        END;

        PROCEDURE ComputeAndDisplay;
        BEGIN
           Fill;
           EnterGraphics;
           Draw;
           ExitGraphics;
        END;

    BEGIN (* Main Program *)
       Hello;
       Initialise;
       ComputeAndDisplay;
       Goodbye;
    END.
```

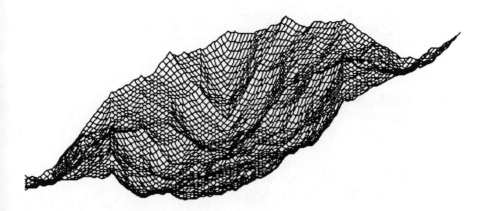

Figure 11.2-1   Fractal mountains.

The theory of graftals is certainly not easy to understand. For those who have not developed their own solution, we here list the appropriate procedure:

Program 11.2-3 (for §8.3)

```
PROCEDURE Graftal;   (* Following Estvanik [1986] *)
    TYPE
        byte = 0..255;
        byteArray = ARRAY[0..15000] OF byte;
        codeArray = ARRAY[0..7, 0..20] OF byte;
        realArray = ARRAY[0..15] OF real;
        stringArray = ARRAY[0..7] OF STRING[20];
    VAR
        code : codeArray;
        graftal : byteArray;
        angle : realArray;
        start : stringArray;
        graftalLength, counter, generationNo, angleNo
                    : integer;
        ready : boolean;

    FUNCTION bitAND (a, b : integer ) : boolean;
        VAR x, y :    RECORD CASE boolean OF
                      False : (counter : integer);
                      True : (seq : SET OF 0..15)
                    END;
        BEGIN
```

```
      x.number := a; y.number := b;
      x.seq : x.seq * y.seq;
      bitAND := x.number <> 0;
  END;   (* of bitAND *)


PROCEDURE codeInput
                    (VAR generationNo : integer;
                  VAR code : codeArray;
                  VAR angle : realArray;
                  VAR angleNo : integer;
                  VAR start : stringArray);
    VAR rule : STRING[20];

    PROCEDURE inputGenerationNo;
    BEGIN
       write (' Generation Number       > ');
       readln (generationNo);
       IF generationNo > 25 THEN generationNo := 25;
    END;


    PROCEDURE inputRule;
       VAR ruleNo, alphabet : integer;
    BEGIN
       FOR ruleNo := 0 TO 7 DO
       BEGIN
          write(' Input of ', ruleNo+1, '. Rule   > ');
          readln (rule);
          IF rule = ' ' THEN
             rule := '0';
          code [ruleNo, 0] := length (rule);
          start [ruleNo] := rule;
          FOR alphabet := 1 TO code [ruleNo, 0] DO
          BEGIN
             CASE rule[alphabet] OF
             '0' : code [ruleNo, alphabet] := 0;
             '1' : code [ruleNo, alphabet] := 1;
             '[' : code [ruleNo, alphabet] := 128;
             ']' : code [ruleNo, alphabet] := 64;
             END;
          END;
       END;
    END;
```

```
PROCEDURE inputAngleNo;
   VAR k, i : integer;
BEGIN
   write (' Number for angle          > ');
   readln (angleNo);
   IF angleNo > 15 THEN
      angleNo := 15;
   FOR k := 1 TO angleNo DO
   BEGIN
      write ('Input of ', k:2, '.Angle (Degrees) > ');
      readln (i);
      angle [k-1] := i*3.14159265 / 180.0;
   END;
END;


PROCEDURE controlOutput;
   VAR alphabet : integer;
BEGIN
   writeln;
   writeln;
   writeln;
   writeln (' Control Output for input of code ' );
   writeln (' --------------------------- ');
   FOR alphabet := 0 TO 7 DO
      writeln (alphabet+1 : 4, start [alphabet] : 20);
   END;
BEGIN
   Textmode;
   inputGenerationNo;
   inputRule;
   inputAngleNo;
   controlOutput;
   CarryOn ('Continue : ');
END;   (* Input of code *)


FUNCTION FindNext (p : integer;
              VAR source : byteArray;
              sourceLength : integer) : integer;
   VAR
      found: boolean;
      depth : integer;
```

```
BEGIN
   depth := 0;
   found := False;
   WHILE (p < sourceLength) AND NOT found DO
   BEGIN
      p := p+1;
      IF (depth = 0) AND (source[p] < 2) THEN
      BEGIN
         findNext := source[p];
         found := True;
      END
      ELSE
      IF (depth = 0) AND (bitAND (source[p], 64)) THEN
         BEGIN
            findNext := 1;
            found := True;
         END
      ELSE IF bitAND (source[p], 128) THEN
         BEGIN
            depth := depth + 1
         END
      ELSE IF bitAND (source[p], 64) THEN
         BEGIN
            depth := depth - 1;
         END
      END;
   IF NOT found THEN
      findNext := 1;
END;    (* of findNext *)


PROCEDURE newAddition (b2, b1, b0 : integer;
                VAR row : byteArray;
                VAR code : codeArray;
                VAR rowLength : integer;
                angleNo ; intger;
   VAR ruleNo, i : integer;
BEGIN
   ruleNo := b2 * 4 + b1 * 2 + b0;
   FOR i := 1 TO code [ruleNo, 0] DO
      BEGIN
         rowLength := rowLength + 1;
```

```
            IF (code[ruleNo, i] >= 0) AND
                         (code[ruleNo, i] <= 63) THEN
               row[rowLength] := code[ruleNo, i];
            IF (code[ruleNo, i] = 64) THEN
               row[rowLength] := 64;
            IF (code[ruleNo, i] = 128) THEN
               row[rowLength] := 128 + Random MOD angleNo;
      END;
END;   (* of newAddition *)


PROCEDURE generation (VAR source : byteArray;
                 VAR sourceLength : integer;
                 VAR code : codeArray);
   VAR
      depth, rowLength, alphabet, k : integer;
      b0, b1, b2 : byte;
      stack : ARRAY[0..200] OF integer;
      row : byteArray;
BEGIN
   depth := 0;
   rowLength := 0;
   b2 := 1;
   b1 := 1;
   FOR alphabet := 1 TO sourceLength DO
   BEGIN
      IF source[alphabet] < 2 THEN
      BEGIN
         b2 := b1;
         b1 := source[alphabet];
         b0 := findNext (alphabet, source, sourceLength);
         newAddition (b2, b1, bo, row, code,
                         sourceLength, angleNo);
      END
   ELSE IF bitAND (source[alphabet], 128) THEN
      BEGIN
         rowLength := rowLength + 1;
         row[rowLength] := source[alphabet];
         depth := depth + 1;
         stack[de[th] := b1;
      END
   ELSE IF bitAND ( source[alphabet], 64) THEN
```

```
    BEGIN
        rowLength := rowLength + 1;
        row[rowLength] := source[alphabet];
        b1 := stack[depth];
        depth := depth - 1;
      END;
    END;
    FOR k := 1 TO rowLength DO source[k] := row[k];
    sourceLength := rowLength;
  END;  (* Of generation *)


PROCEDURE drawGeneration (VAR graftal : byteArray;
                          VAR graftalLength : integer;
                          VAR angle : realArray;
                          VAR counter : integer);
    VAR
        arrayra, arrayxp, arrayyp : ARRAY[0..50] OF real;
        ra, dx, dy, xp, yp, length : real;
        alphabet, depth : integer;
BEGIN
    xp := Xscreen / 2; yp := 0; ra := 0;
    depth := 0; length := 5;
    dx := 0; dy := 0;
    FOR alphabet := 1 TO graftalLength DO
    BEGIN
        IF graftal[alphabet] < 2 THEN
        BEGIN
            GoToPoint (round(xp), round(yp));
            DrawLine (round(xp+dx), round(yp+dy));
            xp := xp+dx;
            yp := yp+dy;
        END;
        IF bitAND (graftal[alphabet], 128) THEN
        BEGIN
            depth := depth + 1;
            arrayra[depth] := ra;
            arrayxp[depth] := xp;
            arrayyp[depth] := yp;
            ra := ra+angle[graftal[alphabet] MOD 16];
            dx := sin(ra) * length;
            dy := cos(ra) * length;
```

```
         END;
         IF bitAND (graftal[alphabet], 64) THEN
         BEGIN
             ra := arrayra[depth];
             xp := arrayxp[depth];
             yp := arrayyp[depth];
             depth := depth - 1;
             dx := sin(ra) * length;
             dy := cos(ra) * length;
         END;
      END;
      CarryOn (' ');
   END   (* Of drawGeneration *)

PROCEDURE printGeneration (VAR graftal: byteArray;
                           VAR graftalLength : integer);
   VAR p : integer;
BEGIN
   writeln ('Graftal Length : ', graftalLength : 6);
   FOR p := 1 TO graftalLength DO
   BEGIN
      IF graftal[p] < 2 THEN write(graftal[p] : 1);
      IF bitAND (graftal[p], 128) THEN write ('[');
      IF bitAND (graftal[p], 164) THEN write (']');
   END;
   writeln;
END;    (* Of printGeneration *)

BEGIN
   inputCode (generationNo, code, angle, angleNo, start);
   graftalLength := 1;
   counter := 1;
   graftal[graftalLength] := 1;
   REPEAT
      generation (graftal, graftalLength, code);
      GraphicsMode;
      drawGeneration (graftal, graftalLength, angle,
                              counter);
      (* Save drawing ''Graftal'' *)
      TextMode;
      printGeneration (graftal, graftalLength);
      writeln ('There were ', counter, ' generations');
```

```
        CarryOn (' More? ');
        counter := counter + 1;
        ready := (graftalLength > 8000) OR
                    ( counter > generationNo)
                  OR button;    (* e.g. Keypressed *)
    UNTIL ready;
  END;
(* END: Problem-specific Procedures *)
```

Insert these procedures into the reference program in the appropriate places.

In the declaration part of the procedure, inside the REPEAT loop, two procedures are called which are enclosed in comment brackets. The procedure SaveDrawing provides a facility for automatically saving the picture to disk. Check in your technical manual to see whether this is possible for your machine. You can also write such a procedure yourself. The procedure printGeneration, if required, prints out the graftal as a string on the screen. It uses the form explained in Chapter 8, with the alphabet {0, 1, [,] }.

The input precedure, which we usually use to read in the data for the screen window, is absent this time. Instead, the required data are read in the procedure codeInput, which is part of the procedure Graftal. After starting the program you will get, e.g., the following dialogue:

```
Generation Number                  >10
Input of 1 .Rule  >  0
Input of 2 .Rule  >  1
Input of 3 .Rule  >  0
Input of 4 .Rule  >  1[01]
Input of 5 .Rule  >  0
Input of 6 .Rule  >  00[01]
Input of 7 .Rule  >  0
Input of 8 .Rule  >  0
Number for angle                   >4
Input of 1 .Angle (Degrees)        >-40
Input of 2 .Angle (Degrees)        >40
Input of 3 .Angle (Degrees)        >-30
Input of 4 .Angle (Degrees)        >30
```

```
Control Output for input of code
-----------------------------------
     1              0
     2              1
     3              0
     4              1[01]
     5              0
     6              00[01]
     7              0
     8              0
Continue: hit <RETURN>
```

Figure 11.2-2 Input dialogue for `Graftal`.

What else is there left to say about Chapter 8? In §8.4 on repetitive patterns a series of program fragments were given, but we think you can embed them in the reference program yourself without any difficulty.

11.3 Ready, Steady, Go!

'Ready, steady, go!' After fractals and graftals we return to the beginning of our computer graphics experiments on Feigenbaum diagrams, in particular landscapes, and the remarkable appearance of the Hénon attractor.

The first exercises that should have stimulated you to experiment were given in the form of the following program fragments:

- 2.1-1   `MeaslesValue`, numerical calculation
- 2.1.1-1   `MeaslesIteration`, graphical representation
- 2.1.2-1   `ParabolaAndDiagonal`, graphical iteration
- 2.2-2   `DisplayFeedback`, output of the feedback constant.

You can complete these without difficulty.

In §2.2.1, which dealt with the bifurcation scenario, we made an attempt to treat the k_i-values of the bifurcation points logarithmically, to estimate the Feigenbaum number (see Exercise 2.2.1-2). Because this exercise involves some difficulty, we give a solution here.

Program 11.3-1 (For Exercise 2.2.1-2)

```
PROCEDURE FeigenbaumIteration;
    VAR
        range, i, iDiv, iMod : integer;
        epsilon, kInfinity, population : real;
        delatxPerPixel : real;
    BEGIN
```

```
epsilon  := (ln (10.0) / 100);
kInfinity := 2.57;
Left := 0.0; Right := 400;
FOR range := 0 TO Xscreen DO
   BEGIN
      iDiv := 1 + range DIV 100;
      iMod := range MOD 100;
      Feedback := kInfinity -
            exp((100-iMod)*epsilon)*exp(-iDiv*ln(10.0));
      population := 0.3;
      IF Feedback > 0 THEN BEGIN
         FOR i := 0 to Invisible DO
            population := f(population, Feedback);
         FOR i:= 0 TO Visible DO
         BEGIN
            SetUniversalPoint (range, population);
            population := f(population, Feedback);
         END;
      END;
   END;
END;
```

In addition, the discussion of how one generates a Feigenbaum landscape may have been a little too brief.

Program 11.3-2 (Feigenbaum landscape)
```
(* BEGIN: Problem-specific procedures *)
   FUNCTION f (p, k : real) : real;
   BEGIN
      f := p + k * p * (1-p);
   END;

   PROCEDURE FeigenbaumLandscape;
      CONST
         lineNo := 100;
         garbageLine = 2;
      TYPE
         box = ARRAY[0..Xscreen] OF integer;
      VAR
         pixel, maximalValue : box;
         i, pixelNo : integer;
```

```
    p, k, real;

PROCEDURE initialisePicture;
   VAR
        j : integer;
BEGIN
(* For the Macintosh you must use:        *)
(* pixelNo = Xscreen-lineNo-WindowBorder *)
(* instead of                             *)
( * pixelNo = Xscreen-lineNo              *)
pixelNo = Xscreen-lineNo
FOR j := 0 TO Xscreen DO
   maximalValue[j] := WindowBorder;
                   (*obliterate everything*)
END;


PROCEDURE initialiseRows (i : integer);
   VAR
      j : integer;
BEGIN
   FOR j := 1 TO pixelNo DO
      pixel[j] := 0;   (* clear pixels *)
   k := Right - j*(Right - Left)/ lineNo;
   p := 0.3;   (* leave start value the same *)
END;


PROCEDURE fill (p : real);
   VAR
      j : integer;
BEGIN
   j := trunc((p-Bottom) * pixelNo / (Top - Bottom));
   IF (j >= 0) AND )j <= pixelNo) THEN
      pixel[j] := pixel[j]+1;
END;


PROCEDURE iterate;
   VAR
      j : integer;
BEGIN
   FOR j := 1 TO Invisible DO
      p := f (p,k);
   FOR j := 1 TO Visible DO
```

```
        BEGIN
           fill (p);
           p := f (p,k);
        END;
     END;


     PROCEDURE sort (i : integer);
        VAR
           j, height : integer;
     BEGIN
        FOR j := 1 TO pixelNo DO
        BEGIN
           height := WindowBorder +
              garbageLines * i + factor * pixel[j];
           IF maximalValue[j+i] < height THEN
              maximalValue[j+i] := height;
        END;
     END;


     PROCEDURE draw (i : integer);
        VAR
           j : integer;
     BEGIN
        SetUniversalPoint (0,0);
        FOR j := 1 TO pixelNo +i DO
           DrawUniversalLine (j, maximalValue[j]);
     END;


  BEGIN
     initialisePicture;
     FOR i := 1 TO lineNo DO
        BEGIN
           initialiseRows (i);
           iterate;
           sort (i);
           draw (i);
        END;   (* for i *)
  END;
(* END : Problem-specific procedures *)
```

In Chapter 3 the Program fragments were given so clearly in Pascal notation that they can easily be incorporated into a complete program. Because some of you may perhaps not have seen the picture of the Rössler attractor, we give it here, together with the corresponding program.

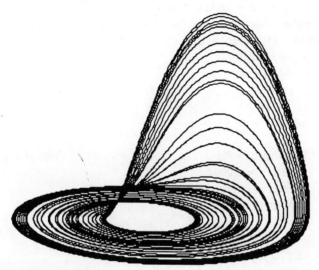

Figure 11.3-1 Rössler attractor.

You obtain this figure if you incorporate the following procedure into your program. Compute 1000 steps without drawing anything, so that we can be sure that the iteration sequence has reached the attractor. Then let the program draw, until we stop it.

Give the variables the following values:

```
Left : = -15; Right := 15;  Bottom := -15; Top := 60;
A := 0.2; B := 0.2;  C := 5.7;
```

Program 11.3-3 (Rössler attractor, see Program Fragment 3.3-1)

```
PROCEDURE Roessler;
  VAR
    i : integer;

  PROCEDURE f;
    CONST
      delta = 0.005;
    VAR
      DX, DY, DZ : REAL;
  BEGIN
    dx := - (y + z);
```

```
            dy := x + y * A;
            dz := B + z * (x - C);
            x := x + delta * dx;
            y := y + delta * dy;
            z := z + delta * dz;
        END;
   BEGIN   (* Roessler *)
        x := -10;
        y := -1;
        z := -1;
        f;
        REPEAT
            f;
            i := i + 1;
        UNTIL i = 1000;
        SetUniversalPoint (x, x+ z + z);
        REPEAT
            f;
            DrawUniversalLine (x, y + z + z);
        UNTIL Button;
   END;   (* Roessler *)
```

'Ready, steady, go' - in this section the speed is often a sprint. This happens because the problems, compared with the Julia and Mandelbrot sets, are very simple and do not require intensive computation. Nevertheless, before we devote our next section to this problem, we will give a few hints for Chapter 4.

The sketches that explain Newton's method must of course be provided with a drawing program, whose central part we now show you:

Program Fragment 11.3-4   (Newton demonstration)

```
   VAR
        N1, N2, N3, StartValue, Left, Right, Top, Bottom : real;
   . . .
   FUNCTION f (x : real) : real;
   BEGIN
        f := (x - N1) * (x - N2) * (x - N3);
   END;


   PROCEDURE drawCurve;
        VAR
            i : integer;
```

```
      deltaX : real;
BEGIN
(* First draw coordinate axes *)
   SetUniversalPoint (Left, 0);
   DrawUniversalLine (Right, 0);
   SetUniversalPoint (0, Bottom);
   DrawUniversalLine (0, Top);
(* Then draw curve *)
   SetUniversalPoint (Left, f(Left));
   i := 0;
   deltaX := (Right - Left) / Xscreen;
   WHILE i <= Xscreen DO
      BEGIN
         DrawUniversalLine (Left+i*deltaX,f(Left+i*deltaX));
         i := i+3;
      END;
END;   (* drawCurve *)


PROCEDURE approximation (x : real);
   CONST
      dx = 0.001;
   VAR
      oldx, fx, fslope : real;
BEGIN (* approximation *)
   REPEAT
      oldx := x;
      fx := f(x);
      fslope := (f(x+dx)-f(x-dx))/(dx+dx)
      IF fslope <> 0.0 THEN
         x := x - fx / fslope
      SetUniversalPoint (oldx, 0);
      DrawUniversalLine (oldx, fx);
      DrawUniversalLine (x, 0);
   UNTIL (ABS(fx) < 1.0E-5);
END;   (* approximation *)


PROCEDURE Initialise;
BEGIN
   ReadReal ('Left          >', Left);
   ReadReal ('Right         >', Right);
   ReadReal ('Bottom        >', Bottom);
   ReadReal ('Top           >', Top);
```

```
   ReadReal ('Root 1         >', N1);
   ReadReal ('Root 2         >', N2);
   ReadReal ('Root 3         >', N3);
   ReadReal ('Start Value    >', StartValue);
END;


PROCEDURE ComputeAndDisplay;
BEGIN
   EnterGraphics;
   drawCurve;
   approximation (StartValue);
   ExitGraphics;
END;
```

In the standard example of Chapter 4 the roots had the values

```
        N1 := -1; N2 := 0; N3 := 1;
```

Of course, that does not prevent you experimenting with other numbers.

11.4   The Loneliness of the Long-distance Reckoner

All our experiments with Julia and Mandelbrot sets have a distressing feature. They take a long time.

In this section we give you some hints for alleviating or exacerbating the loneliness of your long-distance computer.

In Chapter 5 we discussed the representation of Julia sets in the form of a detailed program fragment. We will now provide a full representation as a complete program, in which you see all of the individual procedures combined into a single whole. But again we limit ourselves, as usual, to giving only the problem-specific part and the input procedure.

Program Fragment 11.4-1

```
   PROCEDURE Mapping;
      CONST
         epsq = 0.0025;
      VAR
         xRange, yRange : integer;
         x, y, deltaxPerPixel, deltayPerPixel : real;

      FUNCTION belongsToZa (x, y : real) : boolean;
         CONST
            xa = 1.0;
```

```
          ya = 0.0;
BEGIN
   belongsToZa := (sqr(x-xa)+sqr(y-ya) <= epsq);
END; (* belongsToZa)


FUNCTION belongsToZb (x, y : real) : boolean;
   CONST
       xb = -0.5;
       yb = 0.8660254;
BEGIN
   belongsToZb := (sqr(x-xb)+sqr(y-yb) <= epsq);
END; (* belongsToZb)


FUNCTION belongsToZc (x, y : real) : boolean;
   CONST
       xc = -0.5;
       yc = -0.8660254;
BEGIN
   belongsToZc := (sqr(x-xc)+sqr(y-yc) <= epsq);
END; (* belongsToZc)


FUNCTION JuliaNewtonComputeAndTest (x, y : real)
                  : boolean;
   VAR
      iterationNo : integer;
      finished : boolean;
      xSq, ySq, xTimesy, denominator : real;
      distanceSq, distanceFourth : real;

   PROCEDURE startVariableInitialisation;
   BEGIN
      finished := false;
      iterationNo := 0;
      xSq := sqr(x);
      ySq := sqr(y);
      distanceSq := xSq + ySq;
   END (* startVariableInitialisation *)

   PROCEDURE compute;
   BEGIN
      iterationNo := iterationNo + 1;
```

```
      xTimesy := x*y;
      distanceFourth := sqr(distanceSq);
      denominator := distanceFourth+distanceFourth
                  +distanceFourth;
      x      := 0.666666666*x + (xSq-ySq)/denominator;
      y      := 0.666666666*y -
                (xTimesy+xTimesy)/denominator;
      xSq      := sqr(x);
      ySq      := sqr(y);
      distanceSq := xSq + ySq;
   END;


   PROCEDURE test;
   BEGIN
      finished := (distanceSq < 1.0E-18)
         OR (distanceSq > 1.0E18)
            OR belongsToZa (x,y);
      IF NOT finished THEN finished := belongsToZb (x,y);
      IF NOT finished THEN finished := belongsToZc (x,y);
   END;


   PROCEDURE distinguish;
   BEGIN
   (* Choose one of the statements *)
   (* and delete all the others *)
      JuliaNewtonComputeAndTest :=
         iterationNo = maximalIteration;
      JuliaNewtonComputeAndTest := belongsToZc (x,y)
      JuliaNewtonComputeAndTest :=
         (iterationNo < maximalIteration) AND
                  odd (iterationNo);
      JuliaNewtonComputeAndTest :=
         (iterationNo<maximalIteration) AND
                  (iterationNo MOD 3 = 0);
   END;

BEGIN
   startVariableInitialisation;
   REPEAT
      compute;
      test;
   UNTIL (iterationNo = maxIteration) OR finished;
```

```
              ·    distinguish;
                  END; (* JuliaNewtonComputeAndTest *)

BEGIN
    deltaxPerPixel := (Right - Left ) / Xscreen;
    deltayPerPixel := (Top - Bottom ) / Yscreen;
    y := Bottom;
    FOR yRange := 0 TO yScreen DO
    BEGIN
        x := Left;
        FOR xRange := 0 TO xScreen DO
        BEGIN
            IF JuliaNewtonComputeAndTest (x,y)
                THEN SetUniversalPoint (xRange, yRange);
            x := x + deltaxPerPixel;
        END;
        y := y + deltayPerPixel;
    END;
END; (* Mapping *)

PROCEDURE Initialise;
BEGIN
    ReadReal (' Left          >', Left);
    ReadReal (' Right         >', Right);
    ReadReal (' Bottom        >', Bottom);
    ReadReal (' Top           >', Top);
    ReadInteger ('Maximal Iteration   >', MaximalIteration);
END;
```

And now we give the version for Julia sets using quadratic iteration:

Program Fragment 11.4-2

```
    PROCEDURE Mapping;
        VAR
            xRange, yRange : integer;
            x, y, deltaxPerPixel, delayPerPixel : real;

            FUNCTION JuliaComputeAndTest (x, y : real) : boolean;
                VAR
                    iterationNo : integer;
                    xSq, ySq, distanceSq   : real;
```

```
            finished : boolean;
      PROCEDURE startVariableInitialisation;
      BEGIN
         finished := false;
         iterationNo := 0;
         xSq := sqr(x); ySq := sqr(y);
         distanceSq := xSq + ySq;
      END;  (* startVariableInitialisation *)

      PROCEDURE compute;
      BEGIN
         iterationNo := iterationNo  + 1;
         y := x * y;
         y : = y + y - cImaginary;
         x := xSq - ySq - cReal;
         xSq := sqr(x); ysQ := sqr(y);
         distanceSq := xSq + ySq;
      END;  (* compute *)

      PROCEDURE test;
      BEGIN
         finished := (distanceSq > 100.0);
      END;  (* test *)

      PROCEDURE distinguish;
      BEGIN  (* See also Program Fragment 11.4-1*)
         JuliaComputeAndTest :=
            iterationNo = maximalIteration;
      END;  (* distinguish *)

   BEGIN (* JuliaComputeAndTest *)
      startVariableInitialisation;
      REPEAT
         compute;
         test;
      UNTIL (iterationNo = maximalIteration) OR finished;
      distinguish;
   END; (* JuliaComputeAndTest *)

BEGIN
   deltaxPerPixel := (Right - Left ) / Xscreen;
   deltayPerPixel := (Top - Bottom ) / Yscreen;
```

```
      y := Bottom;
      FOR yRange := 0 TO yScreen DO
      BEGIN
         x := Left;
         FOR xRange := 0 TO xScreen DO
         BEGIN
            IF JuliaComputeAndTest (x,y)
               THEN SetUniversalPoint (xRange, yRange);
            x := x + deltaxPerPixel;
         END;
         y := y + deltayPerPixel;
      END;
   END; (* Mapping *)


   PROCEDURE Initialise;
   BEGIN
      ReadReal (' Left                >', Left);
      ReadReal (' Right               >', Right);
      ReadReal (' Bottom              >', Bottom);
      ReadReal (' Top                 >', Top);
      ReadReal (' cReal               >', cReal);
      ReadReal (' cImaginary          >', cImaginary);
      ReadInteger ('Maximal Iteration  >', MaximalIteration);
   END;
```

We have already explained in Chapter 5 that the wrong choice of c-value can considerably extend the 'loneliness of the long-distance computer' and leave you sittting in front of an empty screen for several hours. To get a quick preview and to shorten the time taken looking for interesting regions, we used the method of backwards iteration. We consider Program Fragments 5.2-3 and 5.2-4 to be so clear that they do not need further elaboration here.

Proceeding from Julia sets, we finally made our 'encounter with the Gingerbread Man'. Again we will collect together the important parts of the program here.

Program Fragment 11.4-3 (see amplifying remarks in Chapter 6)

```
   PROCEDURE Mapping;
      VAR
         xRange, yRange : integer;
         x, y, x0, y0, deltaxPerPixel, deltayPerPixel : real;
```

```
FUNCTION MandelbrotComputeAndTest (cReal, cImaginary
                  : real)
              : boolean;
  VAR
    iterationNo : integer;
    x, y, xSq, ySq, distanceSq : real;
    finished: boolean;

  PROCEDURE  StartVariableInitialisation;
  BEGIN
    finished := false;
    iterationNo := 0;
    x := x0;
    y := y0;
    xSq := sqr(x);
    ySq := sqr(y);
    distanceSq := xSq + ySq;
  END; (* StartVariableInitialisation *)

  PROCEDURE compute;
  BEGIN
    iterationNo := iterationNo + 1;
    y := x*y;
    y := y+y-cImaginary;
    x := xSq - ySq -cReal;
    xSq := sqr(x);
    ySq := sqr(y);
    distanceSq := xSq + ySq;
  END; (* compute *)

  PROCEDURE test;
  BEGIN
    finished := (distanceSq > 100.0);
  END; (* test *)

  PROCEDURE distinguish;
  BEGIN  (* See also Program Fragment 11.4-1 *)
    MandelbrotComputeAndTest : =
        iterationNo = maximalIteration;
  END;  (* distinguish *)
```

```
   BEGIN (* MandelbrotComputeAndTest *)
      StartVariableInitialisation;
      REPEAT
         compute;
         test;
      UNTIL (iterationNo = maximalIteration) OR finished;
      distinguish;
   END;  (* MandelbrotComputeAndTest *)

BEGIN
   deltaxPerPixel := (Right - Left) / Xscreen;
   deltayPerPixel := (Top - Bottom) / Yscreen;
   x0 := 0.0; y0 := 0.0;
   y := Bottom;
   FOR yRange := 0 TO Yscreen DO
   BEGIN
      x:= Left;
      FOR xRange := 0 TO Xscreen DO
         BEGIN
            IF MandelbrotComputeAndTest (x, y)
               THEN SetPoint (xRange, yRange);
            x := x + deltaxPerPixel;
         END;
      y := y + deltayPerPixel;
   END;
END; (*Mapping)

PROCEDURE Initialise;
BEGIN
   ReadReal ('Left                   > '; Left);
   ReadReal ('Right                  > '; Right);
   ReadReal ('Bottom                 > '; Bottom);
   ReadReal ('Top                    > '; Top);
   ReadReal ('MaximalIteration       > '; MaximalIteration );
END;
```

In addition, the five different methods by which we represented the basins of attraction in §6.2 should briefly be mentioned.

The simplest is Case 1. We have already dealt with this in Program Fragment 11.4–3 without saying so. If you give the starting values x0 and y0 in Mapping another value, the computation can explode.

In order that the other four cases can be investigated with the fewest possible program changes, we modify the procedure `Mapping` only slightly. Before calling `MandelbrotComputeAndTest` we insert one from a block of four program segments, which ensure that the right variables change and the others stay constant. The two global variables `FixedValue1` and `FixedValue2` must be read from the keyboard.

Program Fragment 11.4-4 (Cases 2 to 5)

```
PROCEDURE Mapping;
   VAR
      xRange, yRange : integer;
      x, y, x0, y0, deltaxPerPixel, deltayPerPixel : real;
...

BEGIN
   deltaxPerPixel := (Right - Left) / Xscreen;
   deltayPerPixel := (Top - Bottom) / Yscreen;
   y := Bottom;
   FOR yRange := 0 TO Yscreen DO
   BEGIN
      x:= Left;
      FOR xRange := 0 TO Xscreen DO
         BEGIN
            (* Case 2 *)
            x0 := FixedValue1;
            y0 := y;
            cReal := FixedValue2;
            cImaginary := x;
            IF MandelbrotComputeAndTest (x, y)
               THEN SetPoint (xRange, yRange);
            x := x + deltaxPerPixel;
         END;
      y := y + deltayPerPixel;
   END;
END; (*Mapping)
```

```
| (* Case 3 *)                    | (* Case 4 *)                    |
| x0 := FixedValue1;              | x0 := y;                        |
| y0 := y;                        | y0 := FixedValue1;              |
| cReal := x;                     | cReal := FixedValue2;           |
| cImaginary := FixedValue2;      | cImaginary := x;                |
```

```
|  (* Case 5 *)                    |  (* Case 1, alternative *) |
|  x0 := y;                        |  x0 := FixedValue1;        |
|  y0 := FixedValue1;              |  y0 := FixedValue2;        |
|  cReal := x;                     |  cReal := x;               |
|  cImaginary := FixedValue2;      |  cImaginary := y;          |
```

Select whichever version is best for your problem.

```
BEGIN
    ReadReal ('Left              > '; Left);
    ReadReal ('Right             > '; Right);
    ReadReal ('Bottom            > '; Bottom);
    ReadReal ('Top               > '; Top);
    ReadReal ('FixedValue1       > '; FixedValue1 );
    ReadReal ('FixedValue2       > '; FixedValue2 );
    ReadReal ('MaximalIteration  > '; MaximalIteration );
END;
```

The central procedure of the program, with which we generated Figures 6.3-4 to 6.3-6, is given in the next program fragment. The drawing, which almost always takes place within Mapping, is here done inside the procedure computeAndDraw.

Program Fragment 11.4-5 (Quasi-Feigenbaum diagram)

```
PROCEDURE Mapping;
    VAR
        xRange : integer;
        x1, y1, x2, y2, deltaXPerPixel : real;
        dummy : boolean;

    FUNCTION  ComputeAndTest (cReal, cIamginary : real)
                     : boolean;
    VAR
        IterationNo : integer;
        x, y, xSq, ySq, distanceSq : real;
        finished : boolean;

    PROCEDURE StartVariableInitialisation;
    BEGIN
        x := 0.0; y := 0.0;
        finished := false;
        iterationNo := 0;
        xSq := sqr(x); ySq := sqr(y);
```

```
        distanceSq := xSq + ySq;
    END; (* StartVariableInitialisation *)


    PROCEDURE computeAndDraw;
    BEGIN
        iterationNo := iterationNo + 1;
        y := x*y;
        y := y+y-cImaginary;
        x := xSq - ySq - cReal;
        xSq := sqr(x); ySq := sqr(y);
        distanceSq := xSq + ySq;
        IF (iterationNo > bound) THEN
                        SetUniversalPoint (cReal,x);
    END; (* ComputeAndTest *)


    PROCEDURE test;
    BEGIN
        finished := (distanceSq > 100.0);
    END; (* test *)


BEGIN (* ComputeAndTest *)
    StartVariableInitialisation;
    REPEAT
        computeAndDraw;
        test;
    UNTIL (iterationNo = maximalIteration) OR finished;
END (* ComputeAndTest *)


BEGIN
    x1 := 0.1255; y1 := 0. 6503;
    x2 := 0.1098; y2 := 0.882;
    FOR xRange := 0 TO Xscreen DO
        ComputeAndTest
            (x1-(x2-x1)/6+xRange*(x2-x1)/300,
            y1-(y2-y1)/6+xRange*(y2-y1)/300);
END;  (* Mapping *)
```

For §6.4, 'Metamorphoses', in which we dealt with higher powers of complex numbers, we will show you only the procedure Compute. It uses a local procedure compPow. Everything else remains as you have seen it in Program Fragment 11.4-3. Do not forget to give power a reasonable value.

Program Fragment 11.4-6   (High-powered Gingerbread Man)

```
PROCEDURE Compute
   VAR
      t1, t2 : real;

   PROCEDURE compPow (in1r,in1i, power: real;
                      VAR outr, outi: real);
      CONST
         halfpi := 1.570796327;
      VAR
         alpha, r : real;
   BEGIN
      r := sqrt (in1r*in1r + in1i * in1i);
      IF  r > 0.0 then r := exp (power * ln(r));
      IF ABS(in1r) < 1.0E-9 THEN
         BEGIN
            IF in1i > 0.0 THEN alpha := halfpi;
                     ELSE alpha := halfpi + Pi;
         END ELSE BEGIN
            IF in1r > 0.0 THEN alpha := arctan (in1i/in1r)
                     ELSE alpha := arctan (in1i/in1r) + Pi;
         END;
      IF alpha < 0.0 THEN alpha := alpha + 2.0*Pi;
      alpha := alpha * power;
      outr := r * cos(alpha);
      outi := r * sin(alpha);
   END;   (* compPow *)

BEGIN (* Compute *)
   compPow (x, y, power, t1, t2);
   x := t1 - cReal;
   y := t2 - cImaginary;
   xSq := sqr (x);
   ySq := sqr (y);
   iterationNo := iterationNo + 1;
END;   (* Compute *)
```

From Chapter 7 we display only the pseudo-3D representation. This time it was a Julia set that we drew. The remaining program fragments are so clearly listed that it will give you no trouble to include them.

Program Fragment 11.4-7 (Pseudo-3D graphics)

```
TYPE
   D3maxtype = ARRAY[0..Xscreen] OF integer;
VAR
   D3max : D3maxtype;
   Left, Right, Top, Bottom,
   D3factor, CReal, CImaginary : real;
   D3xstep, D3ystep, MaximalIteration, Bound : integer;
   PictureName : STRING;

PROCEDURE D3mapping;
   VAR
      dummy : boolean;
      xRange, yRange : integer;
      x, y, deltaxPerPixel, deltayPerPixel : real;
   FUNCTION D3ComputeAndTest (x, y : real; xRange, yRange
                     : integer)
                     : boolean;
      VAR
         iterationNo : integer;
         xSq, ySq, distanceSq : real;
         finished: boolean;
      PROCEDURE StartVariableInitialisation;
      BEGIN
         finished := false;
         iterationNo := 0;
         xSq := sqr (x);
         ySq := sqr (y);
         distanceSq := xSq + ySq;
      END; (* StartVariableInitialisation *)

      PROCEDURE Compute;   (* Julia-set *)
      BEGIN
         iterationNo := iterationNo + 1;
         y := x * y;
         y := y + y - CImaginary;
         x := xSq - ySq - CReal;
         xSq := sqr (x);
         ySq := sqr (y);
         distanceSq := xSq + ySq;
```

```
      END;   (*Compute *)

      PROCEDURE Test;
      BEGIN
         finished := (distanceSq > 100.0);
      END;    (* Test *)

      PROCEDURE D3set (VAR D3max : D3maxType;
                  column, row, height : integer);
         VAR
            cell, content : integer;
      BEGIN
         cell := column + row - (yScreen -100) DIV 2;
         IF (cell >= 0) AND (cell <= xScreen) THEN
         BEGIN
            content := height * D3factor + row;
            IF content > D3max[cell] THEN
               D3max[cell] := content;
         END;
      END;  (* D3set *)

   BEGIN (* D3ComputeAndTest *)
      D3ComputeAndTest := true;
      StartVariableInitialisation;
      REPEAT
         Compute;
         Test;
      UNTIL (iterationNo = maximalIteration) OR finished;
      D3set (D3max, xRange, yRange, iterationNo);
   END (* D3ComputeAndTest *)

   PROCEDURE D3draw (D3max: D3maxType);
      VAR
         cell, coordinate : integer;
   BEGIN
      setUniversalPoint (Left, Bottom);
      FOR cell := 0 TO xScreen DO
         IF (cell MOD D3xstep = 0) THEN
         BEGIN

(* Warning!  The procedure pensize used below is Macintosh- *)
(* specific and cannot be implemented easily in other       *)
```

```
(* Pascal dialects.  If it does not exist on your computer, *)
(* omit the next few lines of code                          *)
(* from here --------------------------------------------------- *)
                IF cell > 1 THEN
                    IF (D3max[cell] = 100+yRange) AND
                       (D3max[cell-D3xstep]=100+yRange)
                    THEN
                        pensize (1, D3ystep)
                    ELSE
                        pensize (1, 1)
(*------------------------------------------------- to here  *)

                coordinate := D3max[cell];
                IF coordinate >0 THEN DrawLine
                        (cell, coordinate);
            END;
    END;  (* D3draw *)


    BEGIN
        FOR xRange := 0 TO xScreen DO
            D3max[xRange] := 0;
        deltaxPerPixel := (Right - Left) / (xScreen - 100);
        deltayPerPixel := (Top - Bottom) / (yScreen - 100);
        y := Bottom;
        FOR yRange := 0 to (yScreen - 100) DO
            BEGIN
                x := Left;
                FOR xRange := 0 TO (xScreen - 100) DO
                    BEGIN
                        IF (xRange MOD d3ystep = 0) THEN
                            dummy:= D3ComputeAndTest
                                        (x, y, xRange, yRange);
                        x := x + deltaxPerPixel;
                    END;
                D3Draw (D3max);
                y := y + deltayPerPixel;
            END;
    END; (* Mapping *)
    (* END: Problem-specific procedures *)


    PROCEDURE Initialise;
```

```
BEGIN
    ReadReal  ('Left              >', Left);
    ReadReal  ('Right             >', Right);
    ReadReal  ('Top               >', Top);
    ReadReal  ('Bottom            >', Bottom);
    ReadReal  ('c-real            >', CReal);
    ReadReal  ('c-imaginary       >', CImaginary);
    ReadInteger ('Max. iteration No >', MaximalIteration);
    ReadInteger ('3D factor         >', D3factor);
    ReadInteger ('3D step - x       >', D3xstep);
    ReadInteger ('3D step - y       >', D3ystep);
END;
```

11.5 What You See Is What You Get

Now we will eliminate another minor disadvantage in our graphics. Pictures on the screen may appear quite beautiful - but it is of course better to print them out, solving the problem of Christmas and birthday presents, or to record the data directly on floppy disk and reload them into the computer at will.

To make ourselves absolutely clear: we will not tell you here how to produce a screen-dump, so-called *hard copy*, on your computer and with your particular printer. The possible combinations are innumerable, and all the time you find new tricks and tips in the computer magazines. Instead we take the point of view that you have a knob somewhere on your computer which causes whatever is on the screen to be printed out on paper, or that you own a graphics utility program that collects the pictures you have produced in memory or on disk, pretties them up, and prints them.

In this chapter we prefer to go into the problem of *soft copy*, that is, into machine-independent methods for storing information generated by computations. A computer does not have to be capable of graphics under all circumstances: it need only generate the data. Admittedly, it is often possible to draw the corresponding pictures on the same machine. The data obtained in this way can be sent to other chaos researchers, and to other types of computer. From there the output can be processed further, for example as coloured pictures.

And in contrast to all drawing methods that we have explained previously, not a single bit of information is lost, so that we can always distinguish between 'black and white'.

We will explain three methods for storing graphical data. The reason is that speed of operation and compactness of storage often conflict. It is left up to you to select which method to use, if you implement one of the following methods in your programs, or whether you develop an entirely different storage concept. But remember, it must be comprehensible to the people (and their computers) with whom you work. In Figure 11.5-1 we show the three methods together.

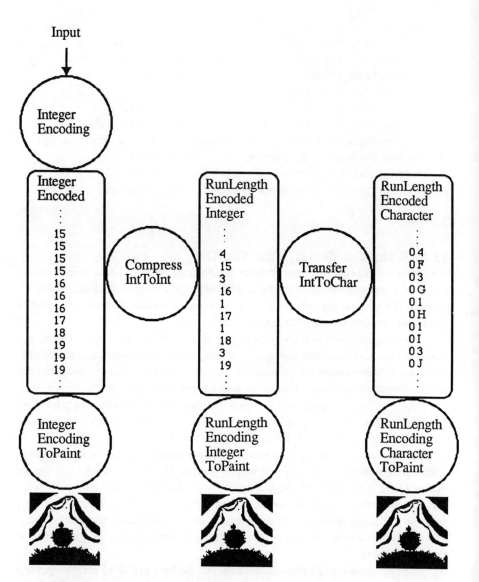

Figure 11.5-1 Three methods of data compression.

The circles represent six programs, which perform the different transformations. Others are certainly possible, and you can develop your own. In the upper part of the picture you can see the different states of the data in our conception. We start with just one idea, in which we communicate with the computer in the form of keyboard input. Instead of producing a graphic (lower picture) straight away, as we have usually done so far, we record the results of the computation as data (long rectangle) on a floppy disk. In the Mapping program this is given by the variable iterationNo, whose current value represents the result of the computation. In order to build in a few possibilities for error detection and further processing, we always write a '0' (which, like negative numbers, cannot occur) when a screen row has been calculated. This number acts as an 'end of line marker'.

The corresponding program (for example the Gingerbread Man) on the whole differs only a little from the version we have seen already. But at the start, instead of the graphics commands, now superfluous, is a variable of type file, and three procedures which make use of it are introduced. These are Store, EnterWriteFile, and ExitWriteFile, which are called in place of Mapping or ComputeAndDisplay. The procedures reset and rewrite are standard Pascal. In some dialects however these may be omitted. This is also true for the implementation of read and write or put and get. If so, you must make special changes to the relevant procedures. Find out about this from your manuals.[1]

The functional procedure MandelbrotComputeAndTest uses, in contrast to the direct drawing program, an integer value, which can immediately be transcribed to disk. We call this method of data representation *integer encoding*.

Program 11.5-1 (Integer encoding)

```
program integerencoding; (* berkeley pascal on sun or vax *)
    const
        stringlength = 8;
        xscreen = 320;  (* e.g. 320 pixels in x-direction *)
        yscreen = 200;  (* e.g. 200 pixels in y-direction *)
    type
        string8 = packed array[1..stringlength] of char;
        intfile = file of integer;
    var
        f : intfile;
        dataname = string8;
        left, right, top, bottom : real;
    (* include further global variables here *)
        maximal iteration : integer;
(* ---------------------- utility ------------------------*)
(* begin: useful subroutines *)
    procedure readstring (information : string8; var value
                          : string8);
```

---

[1] TurboPascal 3.0 running under MS-DOS departs from the language standard. Here reset and rewrite must be used in conjunction with the assign command. In many Pascal implementations the combination of read and write is defined only for text files. In TurboPascal and Berkeley Pascal, however, it is defined more generally.

```
    begin
        write  (information); value := 'pictxxxx';
(* a part for interactive input of the data name in a      *)
(* packed array [...] of char, which we do not give        *)
(* here.  Pascal I/O is often extremely machine-           *)
(* specific, because no data type 'string' is provided.    *)
(*The name here is fixed: pictxxxx (8 letters)             *)
    end;


(* end: useful subroutines *)
(* ---------------------- utility ------------------------ *)


(* ---------------------- file -------------------------- *)
(* begin: file procedures *)
    procedure store (var f: intfile; number : integer);
    begin
      write (f, number);
    end;

    procedure enterwritefile(var f : intfile; filename
                        : string8);
    begin
      rewrite (f, filename);
    end;

    procedure exitwritefile (var f : intfile);
    (* if necessary close(f); *)
    begin;
    end;

(* end: file procedures *)
(* ---------------------- file -------------------------- *)


(* --------------------- application -------------------- *)
(* begin: problem-specific procedures *)

    procedure mapping;
        var
            xrange, yrange : integer;
            x, y, xo, y0, deltaxperpixel
        function mandelbrotcomputeandtest (creal, cimaginary
                        : real)
```

```
                      : integer;
     var
        iterationno : integer;
        x, y, xsq, ysq, distancesq : real;
        finished: boolean;

     procedure  startvariableinitialisation;
     begin
        finished := false;
        iterationno := 0;
        x := x0;
        y := y0;
        xsq := sqr(x);
        ysq := sqr(y);
        distancesq := xsq + ysq;
     end; (* startvariableinitialisation *)

     procedure compute;
     begin
        iterationno := iterationno + 1;
        y := x*y;
        y := y+y-cimaginary;
        x := xsq - ysq -creal;
        xsq := sqr(x);
        ysq := sqr(y);
        distancesq := xsq + ysq;
     end; (* compute *)

     procedure test;
     begin
        finished := (distancesq > 100.0);
     end; (* test *)

     procedure distinguish;
     begin  (* see also program fragment 11.4-1 *)
        mandelbrotcomputeandtest : =iterationno;
     end;   (* distinguish *)

   begin (* mandelbrotcomputeandtest *)
     startvariableinitialisation;
     repeat
```

```
            compute;
            test;
         until (iterationno = maximaliteration) or finished;
         distinguish;
      end;   (* mandelbrotcomputeandtest *)


   begin   (* mapping *)
      deltaxperpixel := (right - left) / xscreen;
      deltayperpixel := (top - bottom) / yscreen;
      x0 := 0.0; y0 := 0.0;
      y := bottom;
      for yrange := 0 to yscreen do
      begin
         x:= left;
         for xrange := 0 to xscreen do
            begin
               store(f,mandelbrotcomputeandtest(x,  y));
               x := x + deltaxperpixel;
            end;
            store (f, 0);      { at the end of each row }
         y := y + deltayperpixel;
      end;
   end; (*mapping)

(* end: problem-specific procedures *)
(* ---------------------- application -------------------- *)

(* ------------------------ main ---------------------- *)
(* begin: procedures of main program *)
   procedure hello;
   begin
      writeln;
      writeln ('computation of picture data ');
      writeln ('------------------------ ');
      writeln; writeln;
   end;

   procedure Initialise;
   begin
      readreal ('left                > '; left);
      readreal ('right               > '; right);
      readreal ('bottom              > '; bottom);
```

```
      readreal ('top                  > '; top);
      readreal ('maximaliteration     > '; maximaliteration );
   end;


   procedure computeandstore;
   begin
      enterwritefile (f, dataname);
      mapping;
      exitwritefile (f);
   end;

(* end: procedures of main program *)
(* ----------------------- main ------------------------ *)
   begin  (* main program *)
      hello;
      initialise;
      computeandstore;
   end.
```

You must be wondering why – in contrast to our own rules of style – everything is written in lower case in this program. The reasons are quite straightforward:

- The program is written in Berkeley Pascal, which is mainly found on UNIX systems with the operating system 4.3BSD. This Pascal compiler accepts only lower case (cf. hints in Chapter 12).
- It is an example to show that you can generate data on any machine in standard Pascal.

This program also runs on a large computer such as a VAX, SUN, or your PC. Only devotees of Turbo Pascal must undertake a small mofidication to the data procedures[2].

A further hint: in standard Pascal the data-type 'string' is not implemented, so that the programmer must work in a very involved manner with the type

```
      packed array [..] of char
```

(cf. procedure readString).

A few other new procedures include a drawing program, which reads the files thus generated and produces graphics from them. The data are opened for reading with reset instead of with rewrite, and Store is replaced by ReadIn. Otherwise MandelbrotComputeAndTest remains the same as before, except that distinguishing what must be drawn occurs within the procedure Mapping.

---

[2]Read the information on the assign command.

Program Fragment 11.5-2 (Integer coding to paint)

```
...
PROCEDURE ReadIn (VAR F : IntFile; VAR number : integer);
BEGIN
   read(F, number);
END;


PROCEDURE EnterReadFile (VAR F: IntFile; fileName
                         : STRING);
BEGIN
   reset (F, fileName);
END;


PROCEDURE ExitReadFile (VAR F: IntFile);
BEGIN
   close (F);
END;


PROCEDURE Mapping;
   VAR
      xRange, yRange, number : intger;
BEGIN
   yRange := 0;
   WHILE NOT EOF (F) DO
   BEGIN
      xRange := 0;
      ReadIn (F, number);
      WHILE NOT (number = 0) DO
      BEGIN
         IF (number = MaximalIteration)
            OR ((number < Bound) AND odd (number)) THEN
               SetPoint (xRange, yRange);
         ReadIn (F, number);
         xRange := xRange + 1;
      END;
      yRange := yRange + 1;
   END;
END;
...
PROCEDURE ComputeAndDisplay;
BEGIN
   EnterGraphics;
```

```
EnterReadFile  (F,  fileName);
Mapping;
ExitReadFile  (F,  fileName);
ExitGraphics;
END;
```

With these two programs we can arrange that what takes a few hours to compute can be drawn in a few minutes. And not only that: if the drawing does not appeal to us, because the contour lines are too close or a detail goes awry, we can quickly produce further drawings from the same data. For this all we need do is change the central IF condition in Mapping. By using

IF (number = MaximalIteration) THEN ...

we draw only the central figure of the Mandelbrot set; but with

IF ((number>Bound) AND (number<maximalIteration)) THEN ...

we draw a thin neighbourhood of it.

This is also the place where we can bring colour into the picture. Depending on the value input for number, we can employ different colours from those available. You will find the necessary codes in your computer manual.

A short mental calculation reveals the disadvantages of integer coding. A standard screen with $320 \times 200 = 64\,000$ pixels requires roughly 128 kilobytes on the disk, because in most Pascal implementations an integer number takes up two bytes of memory. And a larger picture, perhaps in DIN-A 4-format, can easily clog up a hard disk. But the special structure of our pictures gives us a way out. In many regions, nearby points are all coloured the same, having the same iteration depth. Many equal numbers in sequence can be combined into a pair of numbers, in which the first gives the length of the sequence, and the second the colour information. In Figure 11.5-1 you can see this: for example from 15, 15, 15, 15 we get the pair 4, 15. This method is called *run length encoding* and leads to a drastic reduction in storage requirement, to around 20%. That lets us speak of *data compression*.

Because very many disk movements are necessary for this transformation, to do the work on your PC we will use the silent pseudo-disk.[3] This is very quick and more considerate, especially for those family members who are not reminded of the music of the spheres by the singing of a disk-drive motor. And every 'freak' knows that 'it's quicker by RAMdisk'.

A suitable program CompressToInt is given here in standard Pascal (Berkeley Pascal). It uses the file procedures descibed above.

---

[3]Find out how to set up a RAMdisk for your computer.

Program 11.5-3

```pascal
program compresstoint; (* standard pascal *)
const
   stringlength = 8;
type
   intfile = file of integer;
   string8 = packed array[1..stringlength] of char;
var
   dataname : string8;
   in, out: intfile;
   quantity, colour, done : integer;

procedure readin (var f : intfile; var number
                  : integer);
begin
   read(f, number);
end;

procedure enterreadfile (var f: intfile; filename
                  : string8);
begin
   reset (f, filename);
end;

procedure store (var f : intfile; number : integer);
begin
   write (f, number);
end;

procedure enterwritefile (var f : intflie; filename
                  : string8);
begin
   rewrite (f, filename);
end;

procedure exitreadfile (var f: intfile);
   (* if necessary close (f) *)
begin end;

procedure exitwritefile (var f: intfile);
   (* if necessary coose (f) *)
begin end;
```

```
begin
   enterreadfile (in, 'intcoded');
   enterwritefile (out, 'rlintdat');
   while not eof (in) do
      begin
         quantity := 1;
         readin (in, colour);
         repeat
            readin (in, done);
            if (done <> 0) then
               if (done = colour) then
                  quantity := quantity + 1
               else
                  begin
                     store (out, quantity);
                     store (out, colour);
                     colour := done;
                     quantity := 1;
                  end;
         until (done := 0) or eof (in);
         store (out, quantity);
         store (out, number);
         store (out, 0);
      end;
   exitreadfile (in); exitwritefile (out);
end.
```

In the first program, 11.5-1, you can use the procedure readstring to read in an arbitrary data name of length 10. The program requires an input procedure with the name intcoded. The resulting compressed data is always called RLIntDat.

Before the compression run, name your data file as intcoded. Note that standard Pascal requires certain conditions. Always naming the data the same way is, however, not too great a disadvantage. In that way you can automate the entire transformation process – which is useful if you have a time-sharing system, which can process your picture data overnight when the computer is free. You will find hints for this in Chapter 12.

We can also draw the compressed data RLIntDat[4]. For this data type too we show you, in Program Fragment 11.5-4, how to produce drawings. Because we no longer have to specify every individual point, there is a small gain in drawing speed. From 11.5-2 we must change only the procedure Mapping. Whether or not we wish

---

[4]RLIntDat stands for **R**un **L**ength **I**nteger **D**ata.

to represent colour on the screen, we use `DrawLine` to draw a straight line on the screen of the appropriate length, or use `GotoPoint` to go to the appropriate place.

Program Fragment 11.5-4

```
PROCEDURE Mapping;
    VAR
        xRange, yRange, quantity, colour : integer;
BEGIN
    yRange := 0;
    WHILE NOT eof (F) DO
    BEGIN
        xRange := 0;
        GotoPoint (xRange, yRange);
        ReadIn (F, quantity);
        WHILE NOT (quantity = 0) DO
        BEGIN
            xRange := xRange + quantity;
            ReadIn (F, colour);
            IF (colour = MaximalIteration) OR
                ((colour < Bound) AND odd (colour)) THEN
                    DrawLine (xRange -1, yRange)
            ELSE
                GotoPoint (xRange, yRange);
            ReadIn (F, quantity);
        END;
        yRange := yRange + 1;
    END;
END;
```

The third approach, known as the *run length encoded character* method, has advantages and disadvantages compared with the above (RLInt) method. It is more complicated, because it must be encoded before storage and decoded before drawing. However, it uses a genuine Pascal text file. We can change this with special programs (editors) and - what is often more important - transmit it by electronic mail.

In text data there are 256 possible symbols. Other symbols, such as accents, are not uniquely defined in ASCII, so we omit them as far as possible. In fact, we restrict ourselves to just 64 symbols,[5] namely the digits, capital and lower-case letters, and also '>' and '?'.

By combining any pair of symbols we can produce an integer code[6] between 0

---

[5]The method goes back to an idea of Dr Georg Heygster of the regional computer centre of the University of Bremen.

[6]In this way we can represent 4096 different colours. If you need even more, just combine three of the symbols together, to get 262 144 colours. Good enough?

and 4095 (= 64*64-1).   This range of values should not be exceeded by the length information (maximal value is the row length) nor by the colour information (maximal value is the iteration depth).

Program 11.5-5   (Transfer int to char[7])

```
program transferinttochar;
   const stringlength = 8;
   type
      intfile = file of integer;
      charfile = text;
      string8 = packed array[1..stringlength] of char;
   var
      in : intfile;
      outtext : charfile;
      quantity, colour : integer;
      chartable : array[0..63] of char;
      dataname : string8;

   procedure readin (var f : intfile; var quantity
                        : integer);
   begin
      readin (f, quantity);
   end;

   procedure enterreadfile (var f : intfile; filename
                        : string8);
   begin
      reset (f, filename);
   end;

   procedure enterwritefile (var f : charfile; filename
                        : string8);
   begin
      rewrite (f, filename);
   end;

   procedure exitreadfile (var f : intfile);
   begin
      (* if necessary close (f) *)
   end;
```

---

[7]We repeat that in some Pascal compilers read and write must be replaced by put and get, and assignments must be specified for the 'window variable' datavariable.

```
procedure exitwritefile (var f : charfile);
begin
   (* if necessary close (f) *)
end;

procedure store (var outtext : charfile; number
                 : integer);
begin
   if number = 0 then writeln (outtext)
   else
   begin
      write (outtext, chartable[number div 64]);
      write (outtext, chartable[number mod 64]);
   end;
end;

procedure inittable;
   var i : integer;
begin
   for i = 0 to 63 do
   begin
      if i < 10 then
         chartable[i] := chr(ord('0') + i)
      else if i < 36 then
         chartable[i] := chr(ord('0') + i + 7)
      else if i < 62 then
         chartable[i] := chr(ord('0') + i + 13)
      else if i = 62 then
         chartable[i] := '>'
      else if i = 63 then
         chartable[i] := '?';
   end;
end;

begin
   inittable;
   enterreadfile (in, 'rlintdat');
   enterwritefile (outtext, 'rlchardat');
   while not eof (in) do
   begin
```

```
            readin (in, quantity);
            if quantity = 0 then
               store (outtext, 0)
            else
            begin
               store (outtext, quantity);
               readin (in, colour);
               store (outtext, colour);
            end;
         end;
         exitreadfile (in); exitwritefile (outtext);
      end.
```

The coding happens in the two combined programs, 11.5-5 and 11.5-6, by means of a *look-up table*. That is, a table that determines how to encode and decode the appropriate information. These tables are number fields, initialised once and for all at the start of the program. They can then be used for further computations in new runs.

The first of the two programs, 11.5-5, converts the run length encoded integers into characters, and is listed here. Note the line markers, which we insert with writeln (outtext) when a row of the picture has been completed. They make editing easier. We can imagine using them to help carry out particular changes to a picture.

In the final program of this section the drawing will be done. The table contains integer numbers, whose actual characters will be used as an index. The procedure Mapping conforms to the version of 11.5-4: only InitTable must be called at the beginning. The main changes occur in the procedure ReadIn.

Program Fragment 11.5-6 (Run length encoding char to paint)

```
   ...
   TYPE
      CharFile : text;
   VAR
      InText : CharFile;
      IntTable : ARRAY['0'..'z'] OF integer;
   ...
   PROCEDURE InitTable;
      VAR
         ch : char;
   BEGIN
      FOR ch := '0' TO 'z' DO
         BEGIN
            IF ch IN ['0'..'9'] THEN
               IntTable[ch] := ord (ch) - ord ('0')
```

```
          ELSE IF ch IN ['A'..'Z'] THEN
            IntTable[ch] := ord (ch) - ord ('0') - 7
          ELSE IF ch IN ['a'..'z'] THEN
            IntTable[ch] := ord (ch) - ord ('0') - 13
          ELSE IF ch = '>' THEN
            IntTable[ch] := 62
          ELSE IF ch = '?' THEN
            IntTable[ch] := 63
          ELSE
            IntTable[ch] := 0;
    END;
END;


PROCEDURE ReadIn (VAR InText : CharFile; VAR number
                  : integer);
    VAR ch1, ch2 : char;
BEGIN
    IF eoln (InText) THEN

    BEGIN
      readln (InText);
      number := 0;
    END
    ELSE
    BEGIN
      read (InText, ch1);
      read (InText, ch2);
      number := (64*IntTable[ch1]+IntTable[ch2]);
    END;
END;


PROCEDURE Mapping;
    VAR xRange, yRange, quantity, colour : integer;
BEGIN
    yRange := 0;
    WHILE NOT eof (InText) DO
    BEGIN
      xRange := 0;
      GotoPoint (xRange, yRange);
      ReadIn (InText, quantity);
      WHILE NOT (quantity = 0) DO
```

```
      BEGIN
        xRange := xRange + quantity;
        ReadIn (InText, colour);
        IF (colour >= MaximalIteration) OR
           ((colour < Bound) AND odd (colour)) THEN
              DrawLine (xRange - 1, yRange)
        ELSE
           GotoPoint (xRange, yRange);
        ReadIn (InText, quantity);
      END;
      yRange := yRange + 1;
   END;
END;
```

The character files can take up the same space as the compressed integer data - in the above example around 28 kilobytes. On some computers, the storage of an integer number can take up more space than that of a character, implying some saving.

If you transmit the character data using a 300 baud modem, it can take longer than 15 minutes, so you should only do this on local phone lines. We now explain one way to reduce the telephone bill. The text files produced with Program 11.5-5 contain a very variable frequency of individual characters. For example there are very few zeros. In this case the *Huffman method* of text compression (Streichert 1987) can lead to a saving in space of around 50%. The same applies to the telephone bill!

11.6   A Picture Takes a Trip

We can make use of the possibilities described in the previous chapter for screen- and machine-independent generation of data, when we want to send our pictures to a like-minded 'Gingerbread Man investigator'. For this purpose there are basically two methods:

● the normal mail service
● electronic mail ('e-mail').

There is no probem when two experimenters possess the same make of computer. Then it is quite straightforward to exchange programs and data on floppy disks using the normal mail. But we often encounter a situation where one of them has, say, a Macintosh or Atari, and the other has an MS-DOS machine.

In this case one possibility is to connect both computers by cable over the V24 interface, and to transmit programs and data using special software. Many such *file transfer programs* are available for many different computers. We recommend the popular 'Kermit' program, which is available for most machines.[8]

A much simpler, but more expensive, possibility is to send your programs and data

---

[8]Read the hints in §12.7 and the instructions in the Kermit documentation for your computer, to see how to install this.

by telephone, locally or worldwide.

If you live in the same town all you need next to your computer is an acoustic coupler or a modem. In this way you can transmit Pascal programs and picture data between different computers without moving your computer from the desk. And you do not have to mess about with the pin–connections of the V24 interface on different computers, because when you bought the modem for your machine you naturally also bought the appropriate connecting cable.

It is well known how to transmit data over the local telephone network by modem. Not so well known is the fact that worldwide communication networks exist, which can be used to send mail. Of course they are not free.

Basically, by paying the appropriate user fees, anyone can send a letter to – for instance – the USA. Standard communications networks, which can be used from Europe, are CompuServe and Delphi. Of course, the user who wishes to send a picture or a Pascal program must have access to the appropriate network. Another possibility is to send mail over the worldwide academic research network. Only universities and other research institutions can do this.

How do we send e-mail? Easy!

With an acoustic coupler or modem we call the data transmission service of the Post Office and dial the number for the computer to which we wish to send the mail.

First, you must call the DATEX node computer of the German Federal Post[9], and give your NUI[10].

```
DATEX-P: 44 4000 99132
nui dxyz1234
DATEX-P: Password
xxxxxx


DATEX-P: Usercode dxyz1234 active
set 2:0, 3:0, 4:4, 126:0
```

After setting the PAD parameters (so that the input is not echoed and the password remains invisible) then the telephone number of the computer with which we wish to communicate is entered (invisibly).

```
(001) (n, Usercode dxyz1234, packet-length: 128)

RZ Unix system 4.2 BSD
login: kalle
```

---

[9]*Translator's note:* This is the procedure in Germany. In other countries it is very similar, but the names of the services and their telephone numbers are different.

[10]Network User Identification. This consists of two parts: your visible identification and a secret password.

```
Password:
Last login Sat Jul 11 18:09:03 on ttyh3
4.2 BSD UNIX Release 3.07 #3 (root$FBinf) Wed Apr 29 18:12:35
EET 1987
You have mail.
TERM = (vt100)
From ABC007$PORTLAND.BITNET Sat Jul 11 18:53:31 1987
From ABC007$PORTLAND.BITNET Sun Jul 12 01:02:24 1987
From ABC007$PORTLAND.BITNET Sun Jul 12 07:14:31 1987
From ABC007$PORTLAND.BITNET Mon Jul 13 16:10:00 1987
From ABC007$PORTLAND.BITNET Tue Jul 14 03:38:24 1987
kalle$FBinf 1) mail
Mail version 2.18 5/19/83.  Type ? for help.
''/use/spool/mail/kalle'': 5 messages 5 new
>N    1  ABC007$PORTLAND.BITNET Sat  Jul  11  18:53:31  15/534
''Saturday''
From  ABC007$PORTLAND.BITNET  Sun  Jul  12  01:02:24  31/1324
''request''
From ABC007$PORTLAND.BITNET Sun Jul 12 07:14:31 47/2548 ''FT''
From ABC007$PORTLAND.BITNET Mon Jul 13 16:10:00 22/807 ''Auto''
From ABC007$PORTLAND.BITNET Tue Jul 14 03:38:24 32/1362
&2 Message 2:
From ABC007$PORTLAND.BITNET Sun Jul 12 01:02:24 1987
Received: by FBinf.UUCP; Sun, 12 Jul 87 01:02:19 +0200; AA04029
Message-Id: <8707112302.AA04029$FBinf.UUCP>
Received:  by FBinf.BITNET from portland.bitnet(mailer) with bsmtp
Received:  by PORTLAND (Mailer X1.24) id 6622; Sat, 11 Jul 87
19:01:54 EDT
Subject: request
From: ABC007$PORTLAND.BITNET
To: KALLE$RZA01.BITNET
Date: Sat, 11 Jul 87 19:00:08 EDT
Status: R

Dear Karl-Heinz,
it is atrociously hot here this summer.  Instead of sitting on the
deck with a six-pack, i've got to spend the next few weeks putting
together a small survey of information technology.  It has to be
at an elementary level.   In particular I want to include aspects
that are relatively new - such as the connection between computer
graphics and experimental mathematics.
```

```
The simplest example would be a Feigenbaum diagram.
Please send me the Pascal program to gienerate the picture - and,
for safety, a copy of the picture itself - as soon as possible.
By e-mail, the ordinary mail is dreadful as usual and it takes
about 12 days.
Apart from that, there;s nothing much happening here.
When are you coming over next?
Best wishes
Otmar
    &q
Held 5 messages in usr/spool/mail/kalle
0.9u 1.2s 4:54 0% 8t+4d=13<18 19i+28o 38f+110r 0w
kalle$FBinf 2) logout
```

After reading all the news the connection is broken, and we use logout to leave the
Unix system.   The features of Unix and the details of e-mail will not be given here,
because they can differ from one computer to the next. For more information, read your
manual.
    Three days later...

```
DATEX-P: 44 4000 49632
nui dxyz1234
DATEX-P: Password
XXXXXX

DATEX-P: Usercode dxyz1234 active
set 2:0, 3:0, 4:4, 126:0
(001) (n, Usercode dxyz1234, packet-length: 128)

RZ Unix system 4.2 BSD

login: kalle
Password:
Last login Tue Jul 14 07:05:47 on ttyh3
4.2 BSD UNIX Release 3.07 #3 root$FBinf) Wed Apr 29 18:12:35
EET 1987
You have mail.
TERM = (vt100)
kalle$FBinf 2) mail ABC007$Portland.bitnet
Subject Pascalprogram for Feigenbaum
Dear Otmar,
many thanks for your last letter.   Unfortunately I didn't have
```

time to reply until today.

Here is the required Turbopascal program for the Mac.

As usual start with the following inputs:

Left = 1.8

Right = 3.0

Bottom = 0

Top = 1.5

```
-----------------------------cut here --------------------
PROGRAM EmptyApplicationShell; (* TurboPascal on Macintosh *)
    USES MemTypes, QuickDraw;
    CONST
        Xscreen = 320; (* e.g. 320 pixels in x-direction *)
        Yscreen = 200; (* e.g. 200 pixels in y-direction *)
            VAR
                PictureName : string;
                Left, Right, Top, Bottom : real;
            (* include other global variables here *)
```

And so on... we will not give the program at full length here: see §12.4 for the complete listing.

```
(* ---------------------MAIN-----------------*)


BEGIN (*Main Program *)
    Hello;
    Initialise;
    ComputeAndDisplay;
    Goodbye;
END
```

That's it!   Good luck with the survey.

Best wishes Karl-Heinz.

```
EOT
ABC007$Portland.Bitnet... Connecting to portland.bitnet...
ABC007$Portland.Bitnet... Sent
File 2119 Enqueued on Link RZB23
Sent file 2119 on link RZB23 to PORTLAND ABC007
From RZB23: MDTNCM147I SENT FILE 1150 (2119)
ON LINK RZOZIB21 TO PORTLAND IXS2
From RZSTU1: MDTVMB147I SENT FILE 1841 (2119)
```

```
ON LINK DEARN TO PORTLAND ABC007
...
```

You can see on the screen how the letter is sent on from station to station. In a few minutes it has reached its destination. A few minutes later the picture is on the way too...

```
kalle$FBinf ) mail ABC007$Portland.Bitnet
Subject: Picture
Dear Otmar,
now I'm sending you the picture you wanted.
(This file must be converted with BinHex 4.0)

:#dpdE@&bFb''#D@N!&19%G338j8!*!%(L!!N!3EE!#3!`2rN!MGrhIrhIphrpeh
hhAIGGpehUP@U9DT9UP99reAr9Ip9rkU3#11GZhIZhEYhL*!)x6!$'pM!$)f!%!)
J!£K!''2q)N!2rL*!$ri#3!rm)N!1!!*!(J%!J!!)%#!##4$P%JJ'3!rKd)NH2&b
a9D''!3&8+''!3J8)L3''!8#[`#r[1#3''!#3#)!!#!#!!!J!L!!L!)J!)J#))SJ
))US!UJ#U!+S!r`$r!2m!r`!4)N5)%5*%L2m!N!2r!*!$!3)%#''!J3)#U!)!!L!
!!2q!N!F)(#,''J!%#'')J8)N')!+S!3+!!!!3+!!!$K%J`$!)''!B#!36i)#''6
!Z3#j!,N!Z3#j!,N!Z3#j!,N!Z3#j!,N!Z3#j!,N!Z3#j!,N!Z3#j!,N
!Z3#j!,N!Z3#j!,N!Z3#j!,N!Z3#j!,N!Z3#j!,N!Z3#j!,N!Z3#j!,N
!Z3#j!,N!Z3#j!,N!Z3#j!,N!Z3#j!,N!Z3#j!,N!Z3#j!,N!Z3#j!,N
!Z3#j!,N!Z3#j!,N!Z3#j!,N!Z3#j!,N!Z3#j!,N!Z3#j!,N!Z3#j!,N
```

We do not list the entire encoded picture data...

```
!Z3#j!,N!Z3#j!,N!Z3#j!,N!Z3#j!,N!Z3#j!,N!Z3#j!,N!Z3#j!,N
!Z3#j!,N!Z3#j!,N!Z3#j!,N!Z3#j!,N!Z3#j!,N!Z3#j!,N!Z3#j!,N
!Z3#j!,N!Z3#j!,N!Z3#j!,N!Z3#j!,N!Z3#j!,N!Z3#j!,N!'H!!!!:

-------------------

As usual, please cut out the bit between the two colons.
Best wishes, Karl-Heinz
```

A day later, back comes the answer.

```
Message 3:
From ABC007$PORTLAND.BITNET Thu Jul 16 03:19:30 1987
Received: by FBinf.UUCP; Thu 16 Jul 87 03:19:27 +0200; AA04597
Message-Id: <8707160119.AA04597$FBinf.UUCP>
Received:  by FBinf.BITNET from portland.bitnet(mailer) with bsmtp
Received:  by PORTLAND (Mailer X1.24) id 5914; Wed, 15 Jul 87
21:11:18 EDT
```

```
Subject: Picture?Pascalprogram
From: ABC007$PORTLAND.BITNET
To: KALLE$RZA01.BITNET
Date: Wed, 15 Jul 87 21:09:05 EDT
Status: R
```

```
Dear Kalle,
the picture has arrived perfectly.   I have converted it without
any problems using binhex.   You might try using packit next time.
The text file was larger than 10K - the resulting paint file was
only 7K!
More next time,
Otmar.
```

A few explanations to end with.

We have printed the dialogue in full to show you how

- to write your programs in such a form that they can be transferred between computers of different types and different disk formats;
- to consider how to set up communications within a single town or internationally;
- to think about the problem of data compression.

Our pen-pal Otmar has already spoken about this last problem. When two computers communicate directly, binary files, that is, pictures, can be transmitted directly. You can do this with Kermit. But it only makes sense to do it when both computers are of the same make, otherwise the picture cannot be displayed.

Between different computers, we can use the intermediate format set out in §11.5, which changes a picture to a text file. If a picture passes between several computers *en route* to its destination, and if the sending and receiving computers are of the same type, we recommend the use of programs that convert a binary file into a hex file (BinHex program). Such programs are available on any computer. There also exist programs that can take a hex file (e.g. generated from a picture) and compress it still further, to make the text file smaller ('Compress', 'PackIt', etc.).

A very well-known method of data compression is the so-called Huffman method. It achieves an astonishing degree of compression. Of course the person with whom you are communicating must also have such a program. Maybe this is an idea for a joint programming project. Consult the appropriate technical literature, in order to implement the algorithm: Huffmann (1952), Mann (1987), Streichert (1987).

We end the problem of picture transmission here; but the problem of different computers will continue to concern us in Chapter 12.

12 Pascal and the Fig-trees

12.1   Some Are More Equal Than Others – Graphics on Other Systems

In this chapter we show you how to generate the same graphic of a Feigenbaum diagram on different computer systems. We have chosen the Feigenbaum diagram because it does not take as long to produce as, say, the Gingerbread Man. We will give a 'Feigenbaum reference program' for a series of types of computer, operating systems, and programming languages. Using it you can see how to embed your algorithms in the appropriate program.

Our reference picture is shown in Figure 12.1-1.

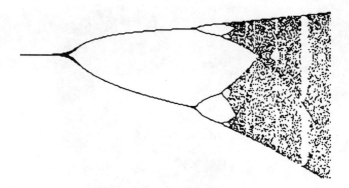

Figure 12.1-1   Feigenbaum reference picture.

12.2   MS-DOS and PS/2 Systems

With IBM's change of direction in 1987, the world of IBM-compatibles and MS-DOS machines changed too. In future as well as MS-DOS there will be a new IBM standard: the OS/2 operating system. The programmer who wishes to develop his Gingerbread Man program on these computers can do so on both families. Turbo Pascal from the Borland company, version 3.0 or higher, is the system of choice. The only difficulties involve different graphics standards and different disk formats, but these should not be too great a problem for the experienced MS-DOS user. The new graphics standard for IBM, like that for the Macintosh II, is a screen of 640 × 480 pixels. Our reference picture has 320 × 200 pixels, corresponding to one of the old standards.

The experienced MS-DOS user will already be aware of the different graphics standards in the world of MS-DOS. It will be harder for beginners, because a program that uses graphics may run on computer X but not on computer Y, even though both are MS-DOS machines. For this reason we have collected here a brief summary of the important graphics standards. In each case check further in the appropriate manual.

MDA (monochrome display adapter)
- 720×348 points
- only text, 9×4 pixels per symbol, only for TTL-monitor

CGA (Colour Graphics Adapter with different modes)
- 00: Text 40×25 monochrome
- 01: Text 40×25 colour
- 02: Text 80×25 monochrome
- 03: Text 80×25 colour
- 04: Graphics 320×200 colour
- 05: Graphics 320×200 monochrome
- 06: Graphics 640×200 monochrome

HGA (Hercules Graphics Adapter)
- 720×348 pixels graphics

EGA (Enhanced Graphics Adapter)
- 640×350 pixels graphics in 16 colours, fore- and background

AGA (Advanced Graphics Adapter)
- Combines the modes of MDA, CGA, HGA

For our reference program we aim at the lowest common denominator: the CGA standard with 320×200 pixels. If you possess a colour screen, you can in this case represent each point in one of four colours. To use the colour graphics commands, see the handbook.

Many IBM-compatible computers use the Hercules graphics card. Unfortunately with this card the incorporation of graphics commands can vary from computer to computer. We restrict ourselves here to the graphics standard defined in the Turbo Pascal handbook from the Borland company for IBM and IBM-compatible computers.

The top left corner of the screen is the coordinate $(0,0)$; x is drawn to the right and y downwards. Anything that lies outside the screen boundaries is ignored. The graphics procedures switch the screen to graphics mode, and the procedure `TextMode` must be called at the end of a graphics program to return the system to text mode.

The standard IBM Graphics card of the old MS-DOS machines up to the Model AT includes three graphics modes. We give here the Turbo Pascal commands and the colour codes:

GraphMode;	GraphColorMode;	HiRes;
320 × 200 pixels	320 × 200 pixels	640 × 200 pixels
$0 \leq x \leq 319$	$0 \leq x \leq 319$	$0 \leq x \leq 639$
$0 \leq y \leq 200$	$0 \leq y \leq 200$	$0 \leq y \leq 199$
black/white	colour	black + one colour

Table 12.2-1 Turbo Pascal commands

Dark	Colours	Light	Colours
0	black	08	dark grey
1	blue	09	light blue
2	green	10	light green
3	cyan	11	light cyan
4	red	12	light red
5	magenta	13	light magenta
6	brown	14	yellow
7	light grey	15	white

Table 12.2-2 Colour codes for high-resolution graphics: Heimsoeth (1985), p. 165.

Table 12.2-2 shows the colour codes needed if you want to use colour graphics. But we recommend you to draw your pictures in black and white. To see that this can produce interesting effects, look at the pictures in this book. The old IBM standard, in our opinion, produces unsatisfying colour pictures: the resolution is too coarse. With the new AGA standard, colour becomes interesting for the user. This of course is also true of colour graphics screens, which can represent 1000 × 1000 pixels with 256 colours. But such screens are rather expensive.

In each of the three graphics modes, Turbo Pascal provides two standard procedures, to draw points or lines:

Plot (x, y, colour); draws in point in the given colour.

Draw (x1, y1, x2, y2, colour): draws a line of the given colour between the specified points.

We will not use any procedures apart from these two.

The graphics routines select colours from a palette. They are called with a parameter between 0 and 3. The active palette contains the currently used colours. That means, for example, that Plot (x, y, colour) with colour = 2 produces red on Palette(0);, with colour = 3 the point is yellow with Palette(2);. Plot (x, y, 0) draws a point in the active background colour. Such a point is invisible. Read the relevant information in the manual.

The following Turbo Pascal reference program is very short, because it works with 'Include-Files'. In translating the main program, two types of data are 'compiled together'. The data `UtilGraph.Pas` contains the useful subroutines and the graphics routines. The data `feigb.Pas` contain the problem-specific part and the input procedure that sets up the data for the Feigenbaum program. We recommend you to construct all of your programs in this way, so that only one command

(*§I feigb.pas *)

relative to the above data names is required.

The Turbo Pascal reference program follows.

Program 12.2-1 (Turbo Pascal reference program for MS-DOS)

```
PROGRAM EmptyApplicationShell;
                        (* only TurboPascal on MS-DOS *)
    CONST
        Xscreen = 320; (* e.g. 320 points in x-direction  *)
        Yscreen = 200; (* e.g. 200 points in y-direction  *)
        palcolour = 1;(*TurboPascal MS-DOS: for palette *)
        dcolour = 15;(*TurboPascal MS-DOS: for draw, plot *)
    TYPE Tstring = string[80];(* only TurboPascal *)
    VAR
        PictureName : Tstring;
        Penx, Peny : integer;
    Left, Right, Top, Bottom : real;
(* Insert further global variables here *)
    Population, Feedback : real;
    Visible, Invisible : integer;


(*$I a:UtilGraph.Pas *)


PROCEDURE Hello;
BEGIN
    ClrScr;
    TextMode;
    InfoOutput ('Representation of                        ');
    InfoOutput ('-----------------------------');
    Newlines (2);
    CarryOn ('Start :');
    Newlines (2);
END;
PROCEDURE Goodbye;
```

```
   BEGIN
      CarryOn ('To finish :');
   END;
(* --------------------------------------------------------- *)
(* Here Include-File with problem-specific procedures        *)
   (*$I a:feigb.Pas *)
(* ------------------------ MAIN ------------------------- *)
   BEGIN
      Hello;
      Initialise;
      ComputeAndDisplay;
      Goodbye;
   END.
```

Two useful hints: In Turbo Pascal there is no predefined data type 'string', having a length of 80. Unfortunately the string length must always be specified in square brackets. Our string variables are thus all of type Tstring. In the Include-Files a: specifies the drive. This is just an example to show what to do when the Turbo Pascal system is on a different drive from the Include-Files.

The Include-File follows: **Util.Graph.Pas**.

```
(* -----------------------UTILITY----------------------- *)
(* BEGIN: Useful subroutines *)
   PROCEDURE ReadReal (information : Tstring; VAR value
                       : real);
   BEGIN
      Write (information);
      ReadLn (value);
   END;


   PROCEDURE ReadInteger (information : Tstring; VAR value
                       : integer);
   BEGIN
      Write (information);
      ReadLn (value);
   END;


   PROCEDURE ReadString (information : Tstring; VAR value :
Tstring);
   BEGIN
      Write (information);
      ReadLn (value);
```

```
      END;

      PROCEDURE InfoOutput (information : Tstring);
      BEGIN
         WriteLn (information);
         WriteLn;
      END;

      PROCEDURE CarryOn (INFORMATION : TSTRING);
      BEGIN
         Write (information, 'Hit <RETURN>');
         ReadLn;
      END;

      PROCEDURE CarryOnIfKey;
      BEGIN
         REPEAT UNTIL KeyPressed;
      END;

      PROCEDURE NewLines (n : integer);
         VAR
            i : integer;
      BEGIN
         FOR i := 0 TO n DO WriteLn;
      END;
(* END: Useful subroutines *)
(* -----------------------UTILITY----------------------- *)

(* -----------------------GRAPHICS-----------------------*)
(* BEGIN:  Graphics Procedures *)

      PROCEDURE SetPoint (xs, ys : integer);
      BEGIN
         (*  Insert machine-specific graphics commands here   *)
         Plot (xs, Yscreen-ys, dcolour);
         Penx := xs; Peny := ys
      END;

      PROCEDURE SetUniversalPoint (xu, yu: real);
         VAR xs, ys : real;
      BEGIN
```

```
      xs := (xu - Left) * Xscreen / (Right - Left);
      ys := (yu - Bottom) * Yscreen / (Top - Bottom);
      SetPoint (round(xs), round(ys));
   END;


   PROCEDURE GoToPoint (xs, ys : integer);
   BEGIN
      Plot (xs, Yscreen - ys, 0);
      Penx := xs; Peny := ys;
   END;


   PROCEDURE DrawLine (xs, ys : integer);
   BEGIN
      Draw (Penx, Yscreen - Peny, xs, Yscreen-ys,
                     dcolour);
      Penx := xs; Peny := ys;
   END;


   PROCEDURE DrawUniversalLine (xu, yu : real);
      VAR xs, ys : real;
   BEGIN
      xs := (xu - Left) * Xscreen/(Right - Left);
      ys := (yu - Bottom) * Yscreen / (Top - Bottom);
      DrawLine (round(xs), round(ys));
   END;


(* PROCEDURE TextMode; implemented in Turbo Pascal *)
(* already exists                                  *)



   PROCEDURE GraphicsMode;
(* DANGER!      DO NOT CONFUSE WITH GraphMode!!!!!  *)
   BEGIN
      ClrScr;
      GraphColorMode;
      Palette (palcolour);
   END;


   PROCEDURE EnterGraphics;
   BEGIN
      Penx := 0;
      Peny := 0;
```

```
      writeln ('To end drawing hit <RETURN> ');
      write ('now hit <RETURN> '); readln;
      Gotoxy(1,23);
      writlen (' ------------ Graphics Mode --------------');
      CarryOn ('Begin :');
      GraphicsMode;
   END;


   PROCEDURE ExitGraphics
(* Machine-specific actions to exit from Graphics Mode        *)
   BEGIN
      ReadLn;
      ClrScr;
      TextMode;
      Gotoxy(1,23);
      Writeln (' -------------- Text Mode ---------------');
   END;

(* END: Graphics Procedures *)
(* -----------------------GRAPHICS-------------------------*)
```

In the implementation of Turbo Pascal for MS-DOS and CP/M computers we must introduce two special global variables Penx and Peny, to store the current screen coordinates.

```
(* ---------------------APPLICATION------------------------*)
(* BEGIN: Problem-specific procedures *)
   FUNCTION f(p, k : real) : real;
   BEGIN f := p + k * p * (1-p);
   END;


   PROCEDURE FeigenbaumIteration;
      VAR
         range, i: integer;
         population, deltaxPerPixel : real;
      BEGIN
         deltaxPerPixel := (Right - Left) / Xscreen;
         FOR range := 0 TO Xscreen DO
            BEGIN
               Feedback := Left + range * deltaxPerPixel;
               population := 0.3;
```

```
            FOR i := 0 TO invisible DO
               population := f(population, Feedback);
            FOR i := 0 TO visible DO
               BEGIN
                  SetUniversalPoint (Feedback, population);
                  population := f(population, feedback);
               END;
         END;
      END;


(* END: Problem-specific procedures                        *)


(*----------------------APPLICATION-----------------------*)

   PROCEDURE Initialise;
   BEGIN
      ReadReal ('Left              >', Left);
      ReadReal ('Right             >', Right);
      ReadReal ('Top               >', Top);
      ReadReal ('Bottom            >', Bottom);
      ReadInteger ('Invisible      >', invisible);
      ReadInteger ('Visible        >', visible);
         (* possibly further inputs *)
      ReadString ('Name of Picture  >', PictureName);
   END;


   PROCEDURE ComputeAndDisplay;
   BEGIN
      EnterGraphics;
      FeigenbaumIteration;
      ExitGraphics;
   END;
```

In our reference program in §11.2 we gave a very clear sequence of procedures. All of these should be retained if you do not work with Include-Files. In the example shown here we have changed these procedures slightly. The procedures also are not grouped according to their logical membership of the class 'Utility', 'Graphics', or 'Main'. Instead, Initialise and ComputeAndDisplay are included among the problem-specific procedures. In this way we can rapidly locate the procedures that must be modified, if another program fragment is used. You need only include new global variables in the main program, and change the initialisation procedure and the procedure calls that lie between EnterGraphics and ExitGraphics.

12.3 UNIX Systems

The word has surely got around by now that UNIX is not just an exotic operating system found only in universities. UNIX is on the march. And so we will give all of you who are in the grip of such an efficient operating system as MS-DOS a few hints on how to dig out the secrets of the Gingerbread Man on UNIX computers.

Let us begin with the most comfortable possibilities. The UNIX system on which you can most easily compute is a SUN or a VAX with a graphics system. UNIX is a very old operating system, and previously people did not think much about graphics. Usually UNIX systems are not equipped with graphics terminals.

If this is the case at your institution, get hold of a compatible standard Pascal compiler (Berkeley Pascal, OMSI Pascal, Oregon Pascal, etc.).   Unfortunately these Pascal systems do not include graphics commands as standard. You must supply the appropropriate commands to set a point on the screen with the help of external C routines. Get together with an experienced UNIX expert, who will quickly be able to write out these C routines, such as setpoint and line. Of course they may already exist.

A typical Pascal program, to represent a picture on a graphics terminal, has the following structure:

```
PROGRAM EmptyApplicationShell;
                    (* Pascal on UNIX systems *)
    CONST
        Xscreen= 320;   (* e.g. 320 points in x-direction *)
        Yscreen= 200;   (* e.g. 200 points in y-direction *)
    VAR
        Left, Right, Top, Bottom : real;
    (* Insert other global variables here *)
        Feedback : real;
        Visible, Invisible : integer;
    PROCEDURE setpoint (x, y, colour : integer);
                    external
    PROCEDURE line (x1, y1, x2, y2, colour : integer);
                    external;

(* ---------------------UTILITY-------------------------- *)
    ...
    BEGIN (* Main Program *)
        Hello;
        Initialise;
        ComputeAndDisplay;
        Goodbye;
    END
```

To incorporate the graphics you can without difficulty use the program fragments from this book. One point that causes problems is the data type 'string', which in standard Pascal is not implemented. Replace the use of this type by procedures you write yourself, using the type `packed array[..] of char`.

Another important hint: in many compilers the declaration of external procedures must occur before all other procedure declarations. In Berkeley Pascal everything must be written in lower case.

UNIX is a multi-user system. On a UNIX system the methods shown here, to calculate the picture data in the computer and display them on the screen, are not entirely fair to other system users. With these methods (see below) you can hang up a terminal for hours.

We recommend that you use the methods of §11.5, generating only the picture data on the UNIX computer, and only after finishing this time-consuming task should you display the results on a graphics screen with the aid of other programs.

This has in particular the consequence that your program to generate the picture data can run as a process with negligible priority in the background. It also allows the possibility of starting such a process in the evening around 10 o'clock, leaving it to run until 7 in the morning, and then letting it 'sleep' until evening, etc. Of course you can also start several jobs one after the other. But remember that all these jobs take up CPU time. Start your jobs at night or at weekends, to avoid arguments with other users or the system manager.

Many UNIX systems do not possess a graphics terminal, but these days they often have intelligent terminals at their disposal. That is, IBM-ATs, Macintoshes, Ataris, or other free-standing personal computers are connected to the UNIX system over a V24 interface at 9600 Baud serial. In this case you should generate your picture data on the UNIX system in one of the three formats suggested (see §11.5) and transfer the resulting data on to floppy disk with a file transfer program. Then you can display the picture calculated on the UNIX system on the screen of your computer.

To make everything concrete, we have collected it together for you here.

The Pascal program for generating the picture data has already been listed in §11.5. After typing Program 11.5-1 into the UNIX system you must compile it and run it. The data can be input interactively. Do not worry if your terminal shows no activity beyond that. Doubtless your program is running and generating picture data. However, it can sometimes happen that your terminal is hung up while the program runs. That can take hours.

Press the two keys $\langle$CTRL$\rangle$ $\langle$Z$\rangle$. This breaks into the program, and the promptline[1] of the UNIX system appears:

%

Type the command `jobs`. The following dialogue ensues:

---

[1]In this case the prompt symbol is the % sign. Which symbol is used differs from shell to shell.

```
% jobs
[1] + Stopped SUNcreate
% bg % 1
[1] SUNcreate
%
```

What does that mean? The square brackets give the operating system reference number of the program that is running. For instance, you can use the `kill` command to cut off this process. The name of the translator program is e.g. `SUNcreate`. By giving the command `bg` and the process number the program is activated and sent into background. The prompt symbol appears and the terminal is free for further work; your job continues to run in the background.[2]

The Pascal program to reconstruct a picture from a text file and diplay it on your computer is also given in §11.5. We recommend you to use the UNIX system as a calculating aid, even if there is no graphics terminal connected to it. You can transfer the resulting picture data to your PC using 'Kermit' and convert them into a picture.

Many users of UNIX systems prefer the programming language C. For this reason we have converted programs 11.5-1, 11.5-3, and 11.5-6 'one to one' into C. Here too the programs should all remain in the background. The resulting data can be read just as well from C as from Pascal.

C Program Examples[3]

Program 12.3-1 (Integer encoded; see 11.5-1)

```
/* EmptyApplicationShell, program 11.5-1 in C */

#include <stdio.h>

#define Xscreen 320 /* e.g. 320 pixels in x-direction */
#define Yscreen 200 /* e.g. 200 pixels in y-direction */
#define Bound 100.0
#define True 1
#define False 0
#define Stringlength 8
typedef char string8 [Stringlength];
typedef FILE *IntFile;
typedef int bool;
    IntFile F;
```

---

[2]This kind of background operation works only in BSD Unix or in Unix versions with a C shell. Ask your system manager, if you want to work in this kind of fashion.

[3]The C programs were written by Roland Meier from the Research Group in Dynamical systems at the University of Bremen.

```
double Left, Right, Top, Bottom;
int MaximalIteration;
/* include further global variables here */
/* ----------------------- file ---------------------- */
/* begin: file procedures */
void Store (F, number)
IntFile F;
int number;
{
   fwrite (&number, sizeof(int), 1, F);
}

void EnterWriteFile(F, Filename)
IntFile *F;
String8 Filename;
{
   *F = fopen(Filename, ''w'');
}

void ExitWriteFile (F)
IntFile *F;
{
   fclose(F);
}

/* end: file procedures */
/* --------------------- file ---------------------- */
/* ------------------- application ------------------ */
/* begin: problem-specific procedures */

double x0, y0;

int MandelbrotComputeAndTest (cReal, cImaginary)
double cReal, cImaginary;
{
#define Sqr(X)  (X)*(X)

      int iterationNo;
      double x, y, xSq, ySq, distanceSq;
      bool finished;
```

```
/* StartVariableInitialisation */
finished = False;
iterationNo = 0;
x = x0;
y = y0;
xSq = sqr(x);
ySq = sqr(y);
distanceSq = xsq + ysq;
do { /* compute */
   iterationNo++;
   y = x*y;
   y = y+y-cImaginary;
   x = xSq - ySq -cReal;
   xSq = sqr(x);
   ySq = sqr(y);
   distanceSq = xsq + ysq;
   /* test */
   finished = (distanceSq > Bound);
} while (iterationNo != MaximalIteration &&
           !finished);
/* distinguish, see also Program 11.5-1 */
return iterationNo;
#undef sqr
}

void Mapping ()
{
   int xRange, yRange;
   double x, y, deltaxPerPixel, deltayPerPixel;

   deltaxPerPixel = (Right - Left) / Xscreen;
   deltay  PerPixel = (Top - Bottom) / Yscreen;
   x0 = 0.0;
   y0 = 0.0;
   y = Bottom;
   for (yRange = 0; yRange < Yscreen; yRange++){
     x= Left;
     for (xRange = 0; xRange < Xscreen; xRange++){
        store (F, MandelbrotComputeAndTest (x, y));
           x += deltaxPerPixel;
        }
```

```
                store (F, 0);      /* at the end of each row */
        y += deltayPerPixel;
    }
}


/* end: problem-specific procedures */
/* -------------------- application ------------------  */

/* ---------------------- main ----------------------- */
/* begin: procedures of main program */
void hello()
{
   printf(''\nComputation of picture data '');
   printf (''\n----------------------- '');
}

void Initialise()
{
   printf (''Left             > ''); scanf(''%lf'', &Left);
   printf (''Right            > ''); scanf(''%lf'', &Right);
   printf ('Bottom          > ''); scanf (''%lf'', &Bottom);
   printf (''Top             > ''); scanf (''%lf'', &Top);
   printf (''Maximal Iteration  > ''); scanf (''%lf'',
                  &MaximalIteration );
/* insert further inputs here */
}

void ComputeAndStore;
{
     Enterwritefile (&F, ''IntCoded'');
     Mapping();
     ExitWriteFile (F);
}


/* end: procedures of main program */
/* ---------------------- main ---------------------- */

main()  /* main program */
   Hello();
   Initialise();
   ComputeAndDisplay();
}
```

Program 12.3.2 (Compress to int; see 11.5-3)

```
/* CompressToInt, Program 11.5-3 in C */
#include <stdio.h>
#define Stringlength  8
typedef FILE *IntFile;
typedef char String8 [Stringlength];

String8 Dataname;
IntFilein, out;
int   quantity, colour, done ;

void EnterReadFile (F, Filename)
IntFile *F;
string8 FileName;
{
   *F = fopen(Filename, ''r'');
}

void EnterWriteFile (F, FileName)
IntFile *F;
string8 FileName;
{
   *F = fopen(FileName, ''w'');
}

void ExitReadFile(F);
IntFile F;
{
   fclose(F);
}

void ExitWritefile(F);
IntFile F;
{
   fclose(F);
}

void ReadIn(F, number)
IntFile F;
int *number;
```

```
{
   fread(number, sizeof(int), 1, F);
}

void store (F, number);
IntFile F;
int number;
{
   fwrite(&b=number, sizeof(int), 1, F);
}

main()
{
   EnterReadFile (&in, ''IntCoded''); /* Integer encoded */
   EnterWriteFile (&out, ''RLIntDat'');
                     /* RL encloded Integer */
   while (!feof(in)) {
      quantity = 1;
      ReadIn (in, &colour);
      if (!feof(in)) {
         do{
            ReadIn (in, &done);
            if (done != 0)
               if (done == colour)
                  quantity++;
               else {
                  store (out, quantity);
                  store (out, colour);
                  colour = done;
                  quantity = 1;
               }
         } while (done != 0 && !feof(in));
         store (out, quantity);
         store (out, number);
         store (out, 0);
      }
   }
   ExitReadFile (in);

   ExitWriteFile (out);
}
```

Program 12.3-3 (Transfer int to char, see 11.5-5)

```
/* TransferIntToChar, Program 11.5-5 in C */

#include <stdio.h>

#define Stringlength 8

typedef FILE*IntFile;
typedef FILE*CharFile;
typedef charString8[Stringlength];
IntFile in;
CharFile outText;
int quantity, colour;
char CharTable[64];
String8 DataName;

void EnterReadFile (F, FileName)
IntFile *F;
String8 FileName;
{
   *F = fopen(Filename, ''r'');
}

void EnterWriteFile (F, FileName)
CharFile *F;
String8 FileName;
{
   *F = fopen(Filename, ''w'');
}

void ExitReadFile (F)
IntFile F;
{
   fclose (F);
}

void Exitwritefile (F)
CharFile F;
{
   fclose (F);
}
```

```c
void ReadIn (F, number)
IntFile F;
int *number;
{
   fread(number, sizeof(int), 1, F);
}


void store (outText, number)
CharFile outText;
int number;
{
   if (number == 0)
      fputc('\n', outText);
   else {
      fputc(CharTable[number/64], outText);
      fputc (CharTable[number % 64], outText);
   }
}


void InitTable()
{
   int i;
      for (i = 0; i < 64; i++) {
         if (i < 10)
            CharTable[i] = '0' + i;
         else if (i < 36)
            CharTable[i] = '0' + i + 7;
         else if (i < 62)
            CharTable[i] = '0' + i + 13;
         else if (i == 62)
            CharTable[i] = '>'
         else if (i == 63)
            CharTable[i] = '?'
   }
}


main()
{
   InitTable();
   EnterReadFile (&in, ''RLIntDat'');
   EnterWriteFile (&outText, ''RLCharDat');
   while (!feof(in)){
```

```
      ReadIn (in, &quantity);
      if (quantity == 0)
         store (outText, 0);
      else {
         store (outText, quantity);
         readin (in, &colour);
         store (outText, colour);
      }
   }
   ExitReadFile (in);
   ExitWriteFfile (outText);
}
```

Those were the hints for UNIX.   UNIX is in particular available on the Macintosh II, which brings us to another operating system and another computer family: the Macintosh.

12.4   Macintosh Systems

There is an enormous range of Pascal implementations for the Macintosh family of computers, all very suitable for computer graphics experiments. They include Turbo Pascal (Borland), Lightspeed Pascal (Think Technologies), TML-Pascal (TML Systems), and MPW (Apple), the Apple development system on the Macintosh. Of course the programs can also be written in other programming languages such as C or Modula II. We now give the corresponding Reference Program for one of the cited implementations.

Program 12.4-1 (Turbo Pascal reference program for Macintosh)

```
PROGRAM EmptyApplicationShell;
                    (* Turbo Pascal on Macintosh *)
USES MemTypes, QuickDraw;
   CONST
      Xscreen = 320;  (*e.g. 320 points in x-direction  *)
      Yscreen = 200;  (*e.g. 200 points in y-direction  *)
   VAR
      PictureName : string;
      Left, Right, Top, Bottom : real;
      (* include additional global variables here *)
      Feedback : real;
      Visible, Invisible : integer;
```

```
(* -----------------------UTILITY--------------------------*)
(* BEGIN:  Useful Subroutines *)
    PROCEDURE ReadReal (information : STRING; VAR value
                          : real);
    BEGIN
      write (information);
      readln (value);
    END;


    PROCEDURE ReadInteger (information : STRING; VAR value
                          : integer);
    BEGIN
      write (information);
      readln (value);
    END;


    PROCEDURE ReadString (information : STRING; VAR value
                          : string);
    BEGIN
      write (information);
      readln (value);
    END;


    PROCEDURE InfoOutput (information : STRING);
    BEGIN
      writeln (information);
      writeln;
    END;


    PROCEDURE CarryOn (information : STRING);
    BEGIN
      write (information, ' hit <RETURN>');
      readln
    END;


    PROCEDURE CarryOnIfKey;
    BEGIN
      REPEAT
      UNTIL KeyPressed
    END;
```

```pascal
   PROCEDURE Newlines (n : integer);
      VAR
         i : integer;
   BEGIN
      FOR i := 1 TO n DO
         writeln;
   END;
```

```pascal
(* END: Useful Subroutines *)
(* ------------------------UTILITY------------------------*)

(* ----------------------GRAPHICS------------------------*)
(* BEGIN:  Graphics Procedures *)

   PROCEDURE SetPoint (xs, ys : integer);
   BEGIN
      (*  Insert machine-specific graphics commands here   *)
      moveto (xs, Yscreen - ys);
      line (0,0)
   END;

   PROCEDURE SetUniversalPoint (xu, yu: real);
      VAR xs, ys : real;
   BEGIN
      xs := (xu - Left) * Xscreen / (Right - Left);
      ys := (yu - Bottom) * Yscreen / (Top - Bottom);
      SetPoint (round(xs), round(ys));
   END;

   PROCEDURE GoToPoint (xs, ys : integer);
   BEGIN
      (*  Insert machine-specific graphics commands here   *)
      moveto (xs, Yscreen - ys);
   END;

   PROCEDURE DrawLine (xs, ys : integer);
   BEGIN
      (*  Insert machine-specific graphics commands here   *)
      lineto (xs, Yscreen - ys);
   END;
```

```
PROCEDURE DrawUniversalLine (xu, yu : real);
  VAR xs, ys : real;
BEGIN
   xs := (xu - Left) * Xscreen/(Right - Left);
   ys := (yu- Bottom) * Yscreen / (Top - Bottom);
   DrawLine (round(xs), round(ys));
END;


PROCEDURE TextMode;
(*  Insert machine-specific graphics commands here      *)
BEGIN
   GotoXY(1,23);
   writlen (' ------------- Text Mode -----------------');
END;


PROCEDURE GraphicsMode;
(*  Insert machine-specific graphics commands here      *)
BEGIN
   ClearScreen;
END;


PROCEDURE EnterGraphics;
BEGIN
   writeln ('To end drawing hit <RETURN> ');
   write ('now hit <RETURN> '); readln;
   GotoXY(1,23);
   writeln('-------------- Graphics Mode ---------------');
   CarryOn('BEGIN :');
   GraphicsMode;
END;


PROCEDURE ExitGraphics
BEGIN
   (*machine-specific actions to exit from Graphics Mode*)
   readln;
   ClearScreen;
   TextMode;
END;

(END: Graphics Procedures *)
(-------------------------GRAPHICS-------------------------*)
```

```
(-----------------------APPLICATION------------------------*)
(BEGIN: Problem-specific procedures *)
(* useful functions for the given application problem        *)

    FUNCTION f(p, k : real) : real;
    BEGIN f := p + k * p * (1-p);
    END;


    PROCEDURE FeigenbaumIteration;
       VAR
          range, i: integer;
          population, deltaxPerPixel : real;
       BEGIN
          deltaxPerPixel := (Right - Left) / Xscreen;
          FOR range := 0 TO Xscreen DO
             BEGIN
                Feedback := Left + range * deltaxPerPixel;
                population := 0.3;
                FOR i := 0 TO Invisible DO
                   population := f(population, Feedback);
                FOR i := 0 TO Visible DO
                   BEGIN
                      SetUniversalPoint (Feedback, population);
                      population := f(population, Feedback);
                   END;
             END;
       END;

(* END: Problem-specific procedures                          *)
(*-----------------------APPLICATION-------------------- *)

    PROCEDURE Initialise;
    BEGIN
       ReadReal  ('Left          >', Left);
       ReadReal  ('Right         >', Right);
       ReadReal  ('Top           >', Top);
       ReadReal  ('Bottom        >', Bottom);
       ReadInteger ('Invisible    >', invisible);
       ReadInteger ('Visible      >', visible);
          (* possibly further inputs *)
       ReadString ('Name of Picture >', PictureName);
    END;
```

```
PROCEDURE ComputeAndDisplay;
BEGIN
   EnterGraphics;
   FeigenbaumIteration;
   ExitGraphics;
END;


(*-------------------------MAIN------------------------- *)
(* BEGIN: Procedures of Main Program *)

   PROCEDURE Hello;
   BEGIN
      TextMode;
      InfoOutput ('Calculation of                      ');
      InfoOutput ('-------------------------');
      Newlines (2);
      CarryOn ('Start :');
      Newlines (2);
   END;

   PROCEDURE Goodbye;
   BEGIN
      CarryOn ('To stop : ');
   END;

(* END: Procedures of Main program *)

(*-------------------------MAIN-------------------------  *)

BEGIN (* Main Program *)
      Hello;
      Initialise;
      ComputeAndDisplay;
      Goodbye;
   END.
```

Of course all of this also works with Include-Files. Compare this program with the reference program from §12.2.

After the start of the program a window appears with the name of the main program. This window is simultaneously a text and graphics window. That is, not only characters but also the usual graphics commands in Turbo Pascal such as ClearScreen and GotoXY apply. It can now be given the appropriate inputs.

On logical grounds we still distinguish here between TextMode and GraphicsMode. In practice we have implemented this in such a way that text called from the procedure TextMode is written in the 22nd row and rolls upwards.

When the drawing is finished, you should see the following picture:

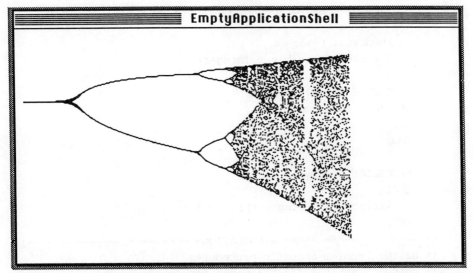

Figure 12.4-1 Turbo Pascal reference picture.

After the drawing is finished the blinking text cursor appears in the left upper corner. You can print out the picture with the key combination

⟨Shift⟩ ⟨Command⟩ ⟨4⟩

or make it into a MacPaint document with

⟨Shift⟩ ⟨Command⟩ ⟨3⟩.

If you then press the ⟨RETURN⟩ key further commands can be given.

Next, some hints on turtle graphics. We described this type of computer graphics experiment in Chapter 8. You can implement your own turtle graphics as in the solutions in §11.3, or rely on the system's own procedures.

In Turbo Pascal on the Macinstosh, as in the MS-DOS version, a turtle graphics library is implemented. Read the description in Appendix F, *Turtle Graphics: Mac graphics made easier* in the *Turbo Pascal Handbook*, editions after 1986.

In addition to Turbo Pascal there is another interesting Pascal implementation. In Lightspeed Pascal it is not possible to modularise into pieces programs that can be compiled with the main program in the form of Include-Files. In this case we recommend that you use the so-called *unit* concept. You can also use units in the Turbo Pascal version. In the main program only the most important parts are given here.

Program 12.4–2 (Lightspeed Pascal reference program for Macintosh)

```
PROGRAM EmptyApplicationShell;
                        (* Lightspeed Pascal Macintosh *)
(* possible graphics library declarations here *)
USES UtilGraph;
(* include further global variables here *)
VAR
   Feedback: real;
   Visible, Invisible : integer;

PROCEDURE Hello;
BEGIN (* as usual *)
...
END;

PROCEDURE Goodbye;
BEGIN
   CarryOn ('To stop : ');
END;
(* --------------------APPLICATION-----------------------*)
(* BEGIN: Problem-specific procedures *)
(* useful functions for the given application problem        *)

FUNCTION f(p, k : real) : real;
BEGIN
   f := p + k * p * (1-p);
END;

PROCEDURE FeigenbaumIteration;
   VAR
      range, i: integer;
      population, deltaxPerPixel : real;
   BEGIN  (* as usual *)
   ...
   END;

PROCEDURE Initialise;
BEGIN  (* as usual *)
...
END;
```

```
    PROCEDURE ComputeAndDisplay;
    BEGIN
       EnterGraphics;
       FeigenbaumIteration;
       ExitGraphics;
    END;
(* END: Problem-specific procedures              *)
(*-----------------------APPLICATION---------------------
    *)

    BEGIN (* Main Program *)
       Hello;
       Initialise;
       ComputeAndDisplay;
       Goodbye;
    END.
```

In the unit UtilGraph we put the data structures and procedures that always remain the same. You can comfortably include all data structures that are global for all programs in such a unit.

```
    UNIT UtilGraph;
    INTERFACE
       CONST
          Xscreen = 320; (* e.g. 320 points in x-direction *)
          Yscreen = 200; (* e.g. 200 points in y-direction *)
       VAR
          CursorShape : CursHandle;
          PictureName : STRING;
          Left, Right, Top. Bottom : real;

    (* ----------------------UTILITY ----------------------- *)
    PROCEDURE ReadReal (information : STRING; VAR value
                        : real);
    PROCEDURE ReadInteger (information : STRING; VAR value
                           : integer);
    PROCEDURE ReadString (information : STRING; VAR value
                          : STRING);
    PROCEDURE InfoOutput (information : STRING);
    PROCEDURE CarryOn (information : STRING);
    PROCEDURE CarryOnIfKey;
```

```
    PROCEDURE Newlines (n : integer);
(* ----------------------UTILITY ----------------------  *)
(* ----------------------GRAPHICS ----------------------  *)
    PROCEDURE InitMyCursor;
    PROCEDURE SetPoint (xs, ys : integer);
    PROCEDURE SetUniversalPoint (xw, yw : real);
    PROCEDURE DrawLine (xs, ys : integer);
    PROCEDURE DrawUniversalLine (xw, yw : real);
    PROCEDURE TextMode;
    PROCEDURE GraphicsMode;
    PROCEDURE EnterGraphics;
    PROCEDURE  ExitGraphics;
(* ----------------------GRAPHICS ----------------------  *)
    IMPLEMENTATION
(* ----------------------UTILITY----------------------------*)
(* BEGIN:  Useful Subroutines *)
    PROCEDURE ReadReal;
    BEGIN
       write (information);
       readln (value);
    END;

    PROCEDURE ReadInteger;
    BEGIN
       write (information);
       readln (value);
    END;

    PROCEDURE ReadString;
    BEGIN
       write (information);
       readln (value);
    END;

    PROCEDURE InfoOutput;
    BEGIN
       writeln (information);
       writeln;
    END;
    PROCEDURE CarryOn;
    BEGIN
       write (information, ' hit <RETURN>');
```

```
      readln;
   END;


   PROCEDURE CarryOnIfKey;
   BEGIN
      REPEAT UNTIL button;   (* Lightspeed Pascal *)
   END;


   PROCEDURE Newlines;
      VAR
         i : integer;
   BEGIN
      FOR i := 1 TO n DO
         writeln;
   END;

(* END: Useful Subroutines *)

(* ------------------------UTILITY-------------------------*)

(* -----------------------GRAPHICS------------------------*)
(* BEGIN:  Graphics Procedures *)

   PROCEDURE InitMyCursor;
   BEGIN
      CursorShape := GetCursor (WatchCursor);
   END;

   PROCEDURE SetPoint (xs, ys : integer);
   BEGIN
      (*  Insert machine-specific graphics commands here   *)
      moveto (xs, Yscreen - ys);
      line (0,0)
   END;

   PROCEDURE SetUniversalPoint (xu, yu: real);
      VAR xs, ys : real;
   BEGIN
      xs := (xu - Left) * Xscreen / (Right - Left);
      ys := (yu - Bottom) * Yscreen / (Top - Bottom);
      SetPoint (round(xs), round(ys));
```

```
END;


PROCEDURE DrawLine (xs, ys : integer);
BEGIN
   lineto (xs, Yscreen - ys);
END;


PROCEDURE DrawUniversalLine (xu, yu : real);
   VAR xs, ys : real;
BEGIN
   xs := (xu - Left) * Xscreen/(Right - Left);
   ys := (yu - Bottom) * Yscreen / (Top - Bottom);
   DrawLine (round(xs), round(ys));
END;


PROCEDURE TextMode;
(*  Insert machine-specific graphics commands here       *)
CONST delta = 50;
VAR window : Rect;
BEGIN
   SetRect (window, delta, delta, Xscreen+delta,
                    Yscreen+delta);
   SetTextRect (window);
   ShowText;  (* Lightspeed Pascal *)
END;


PROCEDURE GraphicsMode;
(*  Insert machine-specific graphics commands here       *)
CONST delta = 50;
VAR window : Rect;
BEGIN
   SetRect (window, delta, delta, Xscreen+delta,
                    Yscreen+delta);
   SetDrawingRect (window);
   ShowDrawing;
   InitMyCursor; (* initialise WatchCursorForm *)
   SetCursor(CursorShape^^); (* set up WatchCursorForm *)
END;


PROCEDURE EnterGraphics;
BEGIN
   writeln ('To end drawing hit <RETURN> ');
```

```
      write ('now hit <RETURN> '); readln;
      writeln('-------------- Graphics Mode --------------');
      CarryOn('BEGIN :');
      GraphicsMode;
   END;

   PROCEDURE ExitGraphics;
   BEGIN
      (*machine-specific actions to exit from Graphics Mode*)
      InitCursor (* call the standard cursor *)
      readln;  (* graphics window no more frozen *)
      SaveDrawing(PictureName);
                      (* store pic as MacPaint document *)
      TextMode;  (* text window appears *)
      writeln('-----------TextMode-------------');
   END;
(* END: Graphics Procedures *)
(* ----------------------GRAPHICS-------------------------*)
   END
```

If you run this program then the following two windows (Figure 12.4-2) appear one after the other.

After inputting the numerical value and pressing <RETURN>, the 'drawing window' of Lightspeed Pascal appears, and the Feigenbaum diagram is drawn. While it is being drawn the 'watch cursor' - the cursor that resembles a wristwatch - is visible. When the drawing is finished the normal cursor appears. Press <RETURN> to give further commands.

As well as Turbo Pascal and Lightspeed Pascal there is a whole series of other Pascal versions or programming languages that will run on the Macintosh. We will make a few remarks about some of these. The graphics procedures are naturally the same in all Pascal versions.

TML Pascal

In contrast to Turbo Pascal and Lightspeed Pascal, TML Pascal generates the most efficient code. Nevertheless we recommend Lightspeed Pascal. In our opinion it is the most elegant development system as regards the simplicity of giving commands to the Macintosh. The graphics procedures are the same, and we do not give a sample program.

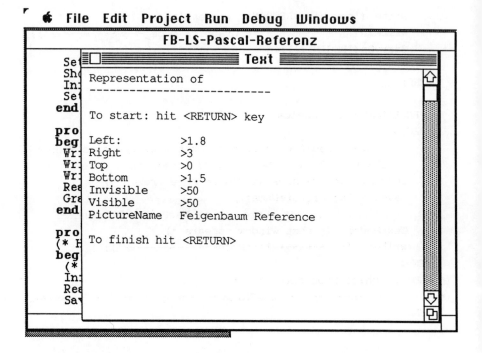

Figure 12.4-2 Screen dialogue.

MPW Pascal

MPW Pascal is a component of the software development environment 'Macintosh Programmer's Workshop', and was developed by Apple. It is based on ANSI Standard Pascal. It contains numerous Apple-specific developments such as the SANE library. SANE complies with the IEEE standard 754 for floating-point arithmetic. In addition MPW Pascal, like TML Pascal, contains facilities for object-oriented programming. MPW Pascal, or Object Pascal, was developed by Apple in conjunction with Niklaus Wirth, founder of the Pascal language.

MacApp

MacApp consists of a collection of object-oriented libraries for implementing the standard Macinstosh user interface. With MacApp you can considerably simplify the development of the standard Macintosh user interface, so that the main parts of Mac programs are at your disposition for use as building blocks. MacApp is a functional (Pascal) program, which can be applied to an individual program for particular extensions and modifications.

Modula II

Another interesting development environment for computer graphics experiments is Modula II. Modula II was developed by Niklaus Wirth as a successor to Pascal. On the Macintosh there are at present a few Modula II versions: TDI-Modula and MacMETH. MacMETH is the Modula II system developed by ETH at Zurich. We choose TDI-Modula here, because the same firm has developed a Modula compiler for the Atari.

Lightspeed C

C holds a special place for all those who wish to write code that resembles machine language. Many large applications have been developed using C, to exploit the portability of assembly programming. Three examples of C programs have been given already in §12.3 (UNIX systems). They were originally programmed in Lightspeed C on the Macintosh and thus run without change on all C compilers.

12.5 Atari Systems

The Atari range has become extremely popular among home computer fans in the last few years. This is doubtless due to the price/power ratio of the machine. Of course, the Atari 1024 is no Macintosh or IBM system 2, but it can produce pretty good Gingerbread Men – in colour.

Among the available programming languages are GFA Basic, ST Pascal Plus, and C. We give our Reference Program here for ST Pascal Plus.

Program 12.5-1 (ST Pascal Plus reference rrogram for Atari)

```
PROGRAM EmptyApplicationShell;
                      (* ST Pascal Plus on Atari *)
    CONST
        Xscreen = 320;   (* e.g. 320 points in x-direction *)
        Yscreen = 200;   (* e.g. 200 points in y-direction *)
(*$I GEMCONST *)
    TYPE Tstring = string[80];
(*$I GEMTYPE *)
    VAR
        PictureName : Tstring;
        Left, Right, Top, Bottom : real;
        (* include additional global variables here *)
        Feedback : real;
        Visible, Invisible : integer;
(*$I GEMSUBS *)
(*$I D:UtilGraph.Pas *)
```

```
    PROCEDURE Hello;
    BEGIN
       Clear_Screen;
       TextMode;
       InfoOutput ('Calculation of                              ');
       InfoOutput ('--------------------------');
       Newlines (2);
       CarryOn ('Start :');
       Newlines (2);
    END;


    PROCEDURE Goodbye;
    BEGIN
       CarryOn ('To stop : ');
    END;

(* ------------------------------------------------------------ *)
(* include file of problem-specific procedures here :------ *)
(*$I D:feigb.Pas *)
(* ----------------------MAIN-------------------------- *)
    BEGIN
       Hello;
       Initialise;
       ComputeAndDisplay;
       Goodbye;
    END.
```

A hint here too for Include-Files: in Include-Files the drive is specified as, for instance, D: . That is just an example to show what to do when the Pascal system is on a different drive from the Include-Files.

The Include-File follows; **UtilGraph.Pas**.

```
(*----------------------------UTILITY---------------------- *)
(* BEGIN:  Useful Subroutines *)
    PROCEDURE ReadReal (information : Tstring; VAR value
                        : real);
    BEGIN
       write (information);
       readln (value);
    END;
```

```
      PROCEDURE ReadInteger (information : Tstring; VAR value
                          : integer);
      BEGIN
         write (information);
         readln (value);
      END;


      PROCEDURE ReadString (information : Tstring; VAR value
                          : string);
      BEGIN
         write (information);
         readln (value);
      END;


      PROCEDURE InfoOutput (information : Tstring);
      BEGIN
         writeln (information);
         writeln;
      END;


      PROCEDURE CarryOn (information : Tstring);
      BEGIN
         write (information, ' hit <RETURN>');
         readln
      END;


      PROCEDURE CarryOnIfKey;
      BEGIN
         REPEAT
         UNTIL KeyPress;    (* NOT as for Turbo! *)
      END;


      PROCEDURE Newlines (n : integer);
         VAR
            i : integer;
      BEGIN
         FOR i := 1 TO n DO
            writeln;
      END;
(* END: Useful Subroutines *)
(* ------------------------UTILITY-----------------------*)
```

```
(* -----------------------GRAPHICS------------------------*)
(* BEGIN:  Graphics Procedures *)

    PROCEDURE SetPoint (xs, ys : integer);
    BEGIN
       (*  Insert machine-specific graphics commands here    *)
       move_to (xs, Yscreen - ys);
       line_to (xs, Yscreen - ys);
    END;

    PROCEDURE SetUniversalPoint (xu, yu: real);
       VAR xs, ys : real;
    BEGIN
       xs := (xu - Left) * Xscreen / (Right - Left);
       ys := (yu - Bottom) * Yscreen / (Top - Bottom);
       SetPoint (round(xs), round(ys));
    END;

    PROCEDURE GoToPoint (xs, ys : integer);
    BEGIN
       (*  Insert machine-specific graphics commands here    *)
       move_to (xs, Yscreen - ys);
    END;

    PROCEDURE DrawLine (xs, ys : integer);
    BEGIN
       (*  Insert machine-specific graphics commands here    *)
       lineto (xs, Yscreen - ys);
    END;

    PROCEDURE GoToUniversalPoint (xu, yu : real);
    BEGIN
       VAR xs, ys : real;
    BEGIN
       xs := (xu - Left) * Xscreen / (Right - Left);
       ys := (yu - Bottom) * Yscreen / (Top - Bottom);
       GotoPoint (round(xs), round(ys));
    END;

    PROCEDURE DrawUniversalLine (xu, yu : real);
       VAR xs, ys : real;
```

```
    BEGIN
        xs := (xu - Left) * Xscreen/(Right - Left);
        ys := (yu - Bottom) * Yscreen / (Top - Bottom);
        DrawLine (round(xs), round(ys));
    END;


    PROCEDURE TextMode;
    BEGIN
        Exit_Gem;
    END;


    PROCEDURE GraphicsMode;
        VAR i : integer
    BEGIN (* machine-specific graphics commands *)
        i := INIT_Gem;
        Clear_Screen;
    END;


    PROCEDURE EnterGraphics;
    BEGIN
        writeln ('To end drawing hit <RETURN> ');
        writeln('-------------- Graphics Mode --------------');
        CarryOn('BEGIN :');
        GraphicsMode;
    END;


    PROCEDURE ExitGraphics
    BEGIN
        (*Machine-specific actions to exit from Graphics Mode*)
        readln;
        Clear_Screen;
        TextMode;
        writeln ('----------------Text Mode ----------------');
    END;

(* END: Graphics Procedures *)
(* ----------------------GRAPHICS----------------------------*)
```

The problem-specific part does not change (see Turbo Pascal, §12.2).

12.6 Apple II Systems

It is certainly possible to take the view that nowadays the Apple IIe is 'getting on a bit'. It is undisputedly slow, but nevertheless everything explained in this book can be achieved on it. The new Apple IIGS has more speed compared with the Apple II, and better colours, so that devotees of the Apple are likely to stay with this range of machines, rather than try to cope with the snags of, e.g., MS-DOS.

Turbo Pascal

Turbo Pascal 3.00A can only run on the Apple II under CP/M. It functions pretty much as in MS-DOS. Specific problems are the graphics routines, which have to be modified for the graphics system of the Apple II. Recently several technical journals (such as *MC* and *c't*) have given hints for this.

TML Pascal/ORCA Pascal

Compared with the Apple IIe the Apple IIGS is a well-tried machine in new clothes. In particular, the colour range is improved. We recommend the use of TML Pascal (see Program 12.6-3) or ORCA Pascal.

UCSD Pascal

Many Pascal devotees know the UCSD system. At the moment there is version 1.3, which runs on the Apple IIe and also the Apple IIGS. Unfortunately until now the UCSD system recognises only 128K of memory, so that extra user memory (on the GS up to 4 MB) can be used only as a RAMdisk.

Basically there are few major changes to our previous reference program. The use of Include-Files or units is possible.[4]

Program 12.6-1 (Reference program for Apple II, UCSD Pascal)

```
PROGRAM EmptyApplicationShell;
                        (* UCSD Pascal on Apple II *)
   USES applestuff, turtlegraphics;
   CONST
      Xscreen = 280;   (* e.g. 280 points in x-direction *)
      Yscreen = 192;   (* e.g. 192 points in y-direction *)
   VAR
      PictureName : string;
      Left, Right, Top, Bottom : real;
      (* include additional global variables here *)
      Feedback : real;
      Visible, Invisible : integer;
```

---

[4]Examples for Include-Files are given in §12.2, for units in §12.4.

```
(*$I Utiltext *)

PROCEDURE Hello;
BEGIN
   TextMode;
   InfoOutput ('Calculation of                          ');
   InfoOutput ('--------------------------');
   Newlines (2);
   CarryOn ('Start :');
   Newlines (2);
END;

PROCEDURE Goodbye;
BEGIN
   CarryOn ('To stop : ');
END;

(* ---------------------------------------------------------- *)
(* include file of problem-specific procedures here :------ *)

(*$I feigb.Pas *)

(* ------------------------MAIN------------------------- *)
BEGIN
   Hello;
   Initialise;
   ComputeAndDisplay;
   Goodbye;
END.
```

Most changes occur in the graphics procedures. UCSD Pascal implements a kind of turtle graphics, which does not distinguish between line and move. Instead the colour of the pen is changed using the procedure pencolor. On leaving a procedure the colour should always be set to pencolor(none). Then a program error, leading to unexpected movements on the graphics screen, cannot harm the picture.

Program 12.6–2 (Include–File of useful subroutines)
```
(*-------------------------UTILITY--------------------- *)
(* BEGIN:  Useful Subroutines *)
   PROCEDURE ReadReal (information : STRING; VAR value
                       : real);
   BEGIN
```

```
    write (information);
    readln (value);
END;


PROCEDURE ReadInteger (information : STRING; VAR value
                       : integer);
BEGIN
  write (information);
  readln (value);
END;


PROCEDURE ReadString (information : STRING; VAR value
                      : string);
BEGIN
  write (information);
  readln (value);
END;


PROCEDURE InfoOutput (information : STRING );
BEGIN
  writeln (information);
  writeln;
END;


PROCEDURE CarryOn (information : STRING );
BEGIN
  write (information, ' hit <RETURN>');
  readln;
END;


PROCEDURE CarryOnIfKey;
BEGIN
  REPEAT
  UNTIL KeyPressed;
END;


PROCEDURE Newlines (n : integer);
  VAR
     i : integer;
BEGIN
  FOR i := 1 TO n DO
```

```
            writeln;
      END;


(* END: Useful Subroutines *)
(* ------------------------UTILITY-------------------------*)
(* ----------------------GRAPHICS-------------------------*)
(* BEGIN:  Graphics Procedures *)

    PROCEDURE SetPoint (xs, ys : integer);
    BEGIN
       (*  Insert machine-specific graphics commands here   *)
       moveto (xs, ys);
       pencolor(white);  move(0);
       pencolor(none);
    END;

    PROCEDURE SetUniversalPoint (xu, yu: real);
       VAR xs, ys : real;
    BEGIN
       xs := (xu - Left) * Xscreen / (Right - Left);
       ys := (yu - Bottom) * Yscreen / (Top - Bottom);
       SetPoint (round(xs), round(ys));
    END;

    PROCEDURE GoToPoint (xs, ys : integer);
    BEGIN
       (*  Insert machine-specific graphics commands here   *)
       moveto (xs, ys);
    END;

    PROCEDURE DrawLine (xs, ys : integer);
    BEGIN
       (*  Insert machine-specific graphics commands here   *)
       pencolor(white);
       moveto (xs, ys);
       pencolor(none);
    END;

    PROCEDURE GoToUniversalPoint (xu, yu : real);
    BEGIN
       VAR xs, ys : real;
    BEGIN
```

```
      xs := (xu - Left) * Xscreen / (Right - Left);
      ys := (yu - Bottom) * Yscreen / (Top - Bottom);
      GotoPoint (round(xs), round(ys));
   END;


   PROCEDURE DrawUniversalLine (xu, yu : real);
      VAR xs, ys : real;
   BEGIN
      xs := (xu - Left) * Xscreen/(Right - Left);
      ys := (yu- Bottom) * Yscreen / (Top - Bottom);
      DrawLine (round(xs), round(ys));
   END;

(* PROCEDURE TextMode is defined in turtle graphics       *)
(* PROCEDURE GraphMode is defined in turtle graphics       *)

   PROCEDURE EnterGraphics;
   BEGIN
      writeln ('To end drawing hit <RETURN> ');
      Page(Output);
      GotoXY(0,0);
      writeln('------------- Graphics Mode --------------');
      CarryOn('BEGIN :');
      InitTurtle;
   END;


   PROCEDURE ExitGraphics
   BEGIN
      (*Machine-specific actions to exit from Graphics Mode*)
      readln;
      TextMode;
      Page(oputput);
      GotoXY(0,0);
   END;

(* END: Graphics Procedures *)
(* -----------------------GRAPHICS-------------------------*)
```

The problem-specific part does not change (see Turbo Pascal, §12.2).
As we have already explained at the start of this chapter, we recommend you to use TML
Pascal on the Apple II. There are two versions:

- TML Pascal (APW)
- TML Pascal (Multi-Window)

The first version is best considered as a version for the software developer who wishes to use Apple Programmer's Workshop, the program development environment from Apple Computer Inc. The second version is easier to use. The multi-window version has a mouse-controlled editor and is similar to TML Pascal on the Macintosh.

Below we once again give our reference program, this time for TML Pascal on the Apple IIGS. For a change, here we draw a Gingerbread Man. We confine ourselves to the most important procedures; you have already seen in previous sections that the surrounding program scarcely changes.

Program 12.6-3 (TML Pascal for Apple IIGS)

```
    PROGRAM EmptyApplicationShell(input, output);
      USES

        ConsoleIO,    (* Library for plain vanilla I/O    *)
        QDIntf;       (* Library for Quick-draw calls     *)
      CONST
        Xscreen = 640.0;
(* SuperHIRES screen 640x400  points*)
        Yscreen = 200;        (* NOTE: real constants!      *)
      VAR
        Left, Right, Top, Bottom : real;
        I, MaximalIteration, Bound : integer;
        R: Rect;


    PROCEDURE ReadReal (s : string; VAR number : real);
    BEGIN
       writeln;
       write (s);
       readln (number);
    END;


    PROCEDURE ReadInteger (s : string; VAR number : integer);
    BEGIN
       writeln;
       write (s);
       readln (number);
    END;


    PROCEDURE SetPoint (x, y, colour : integer);
    BEGIN

       SetDithColor(colour);    (* 16 possible colours *)
```

```
      Moveto (x, y);
      Line (0, 0);
END;


PROCEDURE sqr(x :real) : real;
BEGIN
    sqr (x) := x * x;
END;


PROCEDURE Mapping;
    VAR
        xRange, yRange : intger;
        x, y, x0, y0, deltaxPerPixel, deltayPerPixel : real;


    FUNCTION MandelbrotComputeAndTest (cReal, cImaginary
                        : real)
                 : integer;
    VAR
        iterationNo : integer;
        x, y, xSq, ySq, distanceSq : real;
        finished: boolean;
    BEGIN
(*    StartVariableInitialisation *)
        finished := false;
        iterationNo := 0;
        x := x0;
        y := y0;
        xSq := sqr(x);
        ySq := sqr(y);
        distanceSq := xSq + ySq;
(* StartVariableInitialisation *)
      REPEAT
(* Compute *)
        iterationNo := iterationNo + 1;
        y := x*y;
        y := y+y-cImaginary;
        x := xSq - ySq -cReal;
        xSq := sqr(x); ySq := sqr(y);
        distanceSq := xSq + ySq;
(* Compute *)
(* Test *)
```

```
            finished := distanceSq > 100.0;
(* Test *)
   UNTIL (iterationNo - MaximalIteration) OR finished;
(* distinguish *)
      IF iterationNo = MaximalIteration THEN
         MandelbrotComputeAndTest : = 15
         ELSE BEGIN
            IF iteratioNo > Bound THEN
               MandelbrotComputeAndTest : = 15
            ELSE MandelbrotComputeAndTest : =
                  iterationNo MOD 14;


         END;
(* distinguish *)
      END;
   BEGIN (* Mapping *)
      SetPenSize(1, 1);
      deltaxPerPixel := (Right - Left) / Xscreen;
      deltayPerPixel := (Top - Bottom) / Yscreen;
      x0 := 0.0; y0 := 0.0;
      y := Bottom;
      FOR yRange := 0 TO trunc (yScreen) DO BEGIN
         x:= Left;
         FOR xRange := 0 TO trunc (xScreen) DO BEGIN
            SetPoint (xRange, yRange,
               MandelbrotComputeAndTest (x, y));
            x := x + deltaxPerPixel;
            IF KeyPressed THEN exit  (Mapping);
         END;
         y := y + deltayPerPixel;
      END;
   END;
   PROCEDURE Initialise;
   BEGIN
      ReadReal ('Left               > ', Left);
      ReadReal ('Right              > ', Right);
      ReadReal ('Top                > ', Top);
      ReadReal ('Bottom             > ', Bottom);
      ReadInteger ('Max. Iteration  > ',MaximalIteration);
      ReadInteger ('Bound           > ', Bound);
   END;
```

```
BEGIN (* main *)
   Initialise;
   Mapping;
   REPEAT UNTIL KeyPressed;
END.
```

Observe that:

- We have collected together all local procedures inside the functional procedure MandelbrotComputeAndTest.

- The graphics procedure calls are similar to those for the Macintosh. Therefore it is not necessary for us to list all of them. They are identical.

- The Library ConsoleIO contains the following useful subroutines:
  ```
  Function KeyPressed : boolean
  Procedure EraseScreen;
  Procedure SetDithColor (color : integer);.
  ```

The value for the variable color must lie between 0 and 14, corresponding to the following colours:

Black (0), Dark Grey (1), Brown (2), Purple (3), Blue (4), Dark Green (5), Orange (6), Red (7), Flesh (8), Yellow (9), Green (10), Light Blue (11), Lilac (12), Light Grey (13), White (14).

The way to use colour is easy to see in the program listing. In the algorithmic part distinguish (inside MandelbrotComputeAndTest) the value of IterationNo determines the colour value.

12.7   'Kermit Here' – Communications

In the course of time we become ever more specialised. Now the discussion will be extremely special. The difficult probem of computer-computer communication is the final topic. What we now discuss will be most likely to appeal to 'freaks' – assuming they do not already know it.

The problem is well known: how can we get data and text files from computer X to computer Y? We have already given the main answer in §11.6, 'A Picture Takes a Trip'. But the gap between direct computer-computer connection and e-mail is vast. To bridge it you need to read manuals, think about cables and connectors – all of which takes time. We cannot make the process effortless, but we can give a little help.

First you must buy or assemble a cable, to connect your computer to a modem or another computer via the V24 interface. This hardware problem is the most disagreeable part, because the computer will not do anything if you use the wrong pin-connections. The best solution is the help of a knowledgeable friend, or the purchase of a ready-made cable. Then the next problem raises its head: software. We recommend 'Kermit'. This communications package exists for virtually every computer in the world: try to get this 'public domain' software from usergroups or friends. Of course you can also get other software, provided both computers use the same communications protocol, for instance XModem. Kermit is the most widely available and its documentation is clearly written.

Without going too much into fine detail, we will first describe how file transfer works[5] between an IBM-PC running under MS-DOS and a UNIX system, such as a VAX or SUN.

We assume that your IBM-compatible has two drives or a hard disk, and that the Kermit program is on drive b: or c:. Then you must enter into the following dialogue with the host computer, for example the VAX.

(... means that the screen printout is not given in full.)

Kermit Dialogue: MS-DOS <--> UNIX

```
b>kermit
IBM-PC Kermit-MS V2.26
Type ? for help
Kermit-MS>?
BYE     CLOSE     CONNECT     DEFINE
...
STATUS       TAKE

Kermit-MS>status
Heath-19 emuilation ON          Local Echo Off
...
Communication port: 1           Debug Mode Off
Kermit-MS>set baud 9600
...
```

Now you must set the other parameters such as XON/XOFF, Parity, Stop-bits, 8-bit, etc. as applicable.   Once both computers are similarly configured, the connection can be made.

```
Kermit-MS>connect
[connecting to host, type control-] C to return to PC]
```

Hit <RETURN> and the UNIX system reports...

```
login : kalle
password:
...
```

---

[5]Under Kermit, the procedure is much the same for other machines.

Assume that you want to transfer a file named **rlecdat**, containing your picture data, from the VAX to the PC. Then you do this:

```
VAX>kermit s rlecdat
```

The UNIX computer waits until the PC is ready for the transfer. You must return to the Kermit environment of your running MS-DOS program. Type:

```
<CTRL-]><?>
```
A command line appears. Give it the letter c, as an abbreviation for close connection.

```
COMMAND>c
Kermit-MSreceive picturedata
```

The data rlecdat (on a UNIX system) become picturedata on the MS-DOS system. The transfer then starts, and its successful completion is reported.

```
Kermit-MS>dir
```

The transferred file is written into the directory of your MS-DOS machine. If you already have a file of the same name on your PC, things do not work out very well.

The converse procedure is also simple:

```
VAX>Kermit r
```

You return to the Kermit environment and give the command:

```
Kermit-MS>send example.p
```

The transfer then begins in the opposite direction...

Everything works much the same on other computers, including the Macintosh. But here there is an elegant variant. In the USA there are two programs named MacPut and MacGet, which can transfer Macintosh text and binary files under the MacTerminal 1.1 protocol to and from a UNIX system.

Get hold of the C source files, transfer them to your VAX, compile them into the program – and you are free of all Kermit problems.

```
login: kalle
password:
4.3 BSD UNIX °4: Thu Feb 19 16:00:24 MET 1987
Mon Jul 27 18:28:20 MET DST 1987
SUN>macget msdosref.pas -u
```

```
SUN>ls -1
total 777
-rwxrwxr-x    1 kalle      16384 Jan 16 1987 macget
-rw-r--r--    1 kermit      9193 Jan 16 1987 macget.c.source
-rwxrwxr-x    1 kalle      19456 Jan 16 1987 macget
-rw-r--r--    1 kermit      9577 Jan 16 1987 macget.c.source
-rw-rw-r--    1 kalle       5584 Jul 27 18:33 msdosref.pas
```

13 Appendices

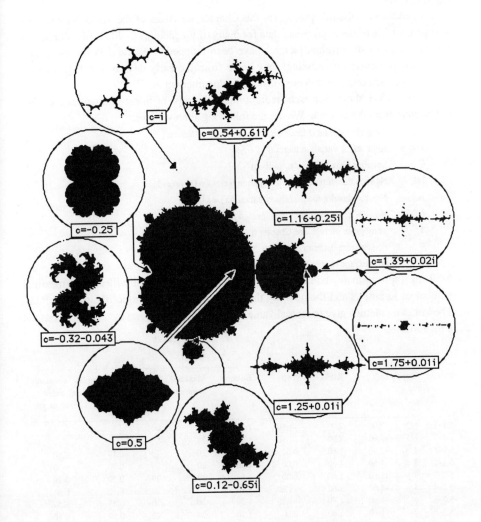

c=i

c=0.54+0.61i

c=-0.25

c=1.16+0.25i

c=1.39+0.02i

c=-0.32-0.043

c=1.75+0.01i

c=0.5

c=1.25+0.01i

c=0.12-0.65i

13.1   Data for Selected Computer Graphics

Especially in your first investigations, it is useful to know where the interesting sections occur. Then you can test out your own programs there. Some of the more interesting Julia sets, together with their parameters in the Mandelbrot set, are collected together below.

In addition to the title picture for this Chapter, an 'Atlas' of the Mandelbrot set, we also give a table of the appropriate data for many of the pictures in this book. For layout reasons many of the original pictures have been cropped. Some data for interesting pictures may perhaps be missing: they come from the early days of our experiments, where there were omissions in our formulas or documentation.

Table 13–1 shows for each figure number the type of picture drawn, as this may not be clear from the caption. We use the following abbreviations:

G	Gingerbread man (Mandelbrot set) or variation thereon
J	Julia set, or variation thereon
C	Set after Curry, Garnett, Sullivan
N1	Newton development of the equation $f(x) = (x-1)*x*(x+1)$
N3	Newton development of the equation $f(x) = x^3-1$
N5	Newton development of the equation $f(x) = x^5-1$
T	Tomogram picture, see Chapter 6
F	Feigenbaum diagram
*	See text for further information

Near the (approximate) boundaries of the picture sections you will see the maximal number of iterations and the quantity that determines the spacing of the contour lines. The last two columns give the initial value for Mandelbrot sets, and the complex constant c for Julia sets.

Picture	Type	Left	Right	Top	Bottom	Maximal iteration	Bound	Complex constant or initial value c_r or x_0	c_i or y_0
5.1–2ff	N3	−2.00	2.00	−1.50	1.50	*	*	–	–
5.1–7	N5	−2.00	2.00	−1.50	1.50	25	5	–	–
5.2–1	J	−1.60	1.60	0.00	1.20	20	20	0.10	0.10
5.2–2	J	−1.60	1.60	0.00	1.20	20	20	0.20	0.20
5.2–3	J	−1.60	1.60	0.00	1.20	20	20	0.30	0.30
5.2–4	J	−1.60	1.60	−1.20	1.20	20	20	0.40	0.40
5.2–5	J	−1.60	1.60	−1.20	1.20	20	20	0.50	0.50
5.2–6	J	−1.60	1.60	0.00	1.20	20	20	0.60	0.60
5.2–7	J	−1.60	1.60	0.00	1.20	20	20	0.70	0.70
5.2–8	J	−1.60	1.60	0.00	1.20	20	20	0.80	0.80
5.2–9	J	−1.75	1.75	−1.20	1.20	200	0	0.745 405 4	0.113 006 3
5.2–10	J	−1.75	1.75	−1.20	1.20	200	0	0.745 428	0.113 009
5.2–11	J	0.15	0.40	−0.322	−0.15	200	40	0.745 405 4	0.113 006 3
5.2–12	J	0.29	0.316 4	−0.209 1	−0.191 4	400	140	0.745 405 4	0.113 006 3
5.2–13	J	0.295 11	0.298 14	−0.203 3	−0.201 3	400	140	0.745 405 4	0.113 006 3

Picture	Type	Left	Right	Top	Bottom	Maximal iteration	Bound	Complex constant or initial value	
								c_r or $x0$	c_i or $y0$
5.2–14	J	0.295 11	0.298 14	−0.2033	−0.2013	400	140	0.745 428 4	0.113 009
5.2–15	J	0.296 26	0.296 86	−0.2024	−0.202	600	200	0.745 405 4	0.113 006 3
5.2–16	J	0.296 26	0.296 86	−0.2024	−0.202	600	200	0.745 428	0.113 009
5.2–17	J	−1.75	1.75	−1.20	1.20	50	0	0.745 428	0.113 009
5.2–18	J	−2.00	1.50	−1.20	1.20	1000	12	0.745 428	0.113 009
5.2–19	J	−0.08	0.07	−0.1	0.1	200	60	0.745 428	0.113 009
5.2–20	J	−0.08	0.07	−0.1	0.1	300	0	0.745 428	0.113 009
5.2–23	J	−1.75	1.75	−1.20	1.20	60	10	1.25	0.011
6.1–1	J	−0.05	0.05	−0.075	0.075	400	150	0.745 405 4	0.113 006 3
6.1–2	J	−0.05	0.05	−0.075	0.075	400	150	0.745 428	0.113 009
6.1–3	G	−1.35	2.65	−1.50	1.50	4	0	0.00	0.00
6.1–4	G	−1.35	2.65	−1.50	1.50	6	0	0.00	0.00
6.1–5	G	−1.35	2.65	−1.50	1.50	8	0	0.00	0.00
6.1–6	G	−1.35	2.65	−1.50	1.50	10	0	0.00	0.00
6.1–7	G	−1.35	2.65	−1.50	1.50	20	0	0.00	0.00
6.1–8	G	−1.35	2.65	−1.50	1.50	100	16	0.00	0.00
6.1–9	G	−1.35	2.65	−1.50	1.50	60	0	0.00	0.00
6.1–10	G	−0.45	−0.25	−0.10	0.10	40	40	0.00	0.00
6.1–11	G	1.934 68	1.949 3	−0.005	0.009	100	20	0.00	0.00
6.1–12	G	0.74	0.75	0.108	0.115 5	120	100	0.00	0.00
6.1–13	G	0.74	0.75	0.115 5	0.123	120	100	0.00	0.00
6.1–14	G	−0.465	−0.45	0.34	0.35	200	60	0.00	0.00
6.2–1ff	T	−2.10	2.10	−2.10	2.10	100	7	*	*
6.2–11	T	0.62	0.64	0.75	0.80	250	100	$y0=0.1$	$c_i=0.4$
6.3–4ff	F	0.60	0.90	0.00	1.50	50	250	–	–
6.4–1	C	−2.50	2.00	−2.00	2.00	250	0	0.00	0.00
6.4–2	C	−0.20	0.40	1.50	1.91	100	0	0.00	0.00
6.4–3ff	G*	−2.10	2.10	−2.10	2.10	100	7	0.00	0.00
6.4–6	C	0.90	1.10	−0.03	0.10	100	0	0.00	0.00
6.4–7	J*	−2.00	2.00	−2.00	2.00	225	8	−0.50	0.44
7.1–1	N1	−2.00	2.00	−1.50	1.50	20	0	–	–
7.1–2	N3	1.00	3.40	−4.50	−2.70	20	0	–	–
7.1–3	J	−2.00	2.00	−1.50	1.50	10	0	0.50	0.50
7.1–4	G	−1.35	2.65	−1.50	1.50	15	0	0.00	0.00
7.1–5	J	−2.00	2.00	−1.50	1.50	20	0	0.745	0.113
7.1–6	G	−1.35	2.65	−1.50	1.50	20	0	0.00	0.00
7.2–1	G	−4.00	1.50	−2.00	2.00	40	12	0.00	0.00
7.2–2	G*	−1.50	1.50	−0.10	1.50	40	7	0.00	0.00
7.2–3	G*	−3.00	3.00	−2.25	2.25	30	10	0.00	0.00
7.2–4	N3	−2.00	2.00	−1.00	1.50	40	3	–	–
7.2–5	J	−2.00	2.00	−1.50	1.50	30	10	1.39	−0.02
7.2–6	J	−18.00	18.00	−13.50	13.50	30	10	1.39	−0.02
7.2–7	J	−2.00	2.00	−1.50	1.50	30	30	−0.35	−0.004
7.2–8	J	−3.20	3.20	−2.00	4.80	30	30	−0.35	−0.004
7.4–1	G	−1.35	2.65	−1.50	1.50	20	0	0.00	0.00
7.4–2	G	−1.35	2.65	−1.50	1.50	20	0	0.00	0.00
7.4–3	J	−2.00	2.00	−1.50	1.50	30	0	0.50	0.50
7.4–4	J	−2.00	2.00	−1.50	1.50	30	5	−0.35	0.15
9.5	J	−1.00	1.00	−1.20	1.20	100	0	−0.30	−0.005
10–1	G	−1.35	2.65	−1.50	1.50	20	0	0.00	0.00

Picture	Type	Left	Right	Top	Bottom	Maximal iteration	Bound	Complex constant or initial value	
								c_r or $x0$	c_i or $y0$
10–2	G	0.80	0.95	–0.35	–0.15	25	0	0.00	0.00
10–3	G	0.80	0.95	–0.35	–0.15	25	15	0.00	0.00
10–4	G	0.85	0.95	–0.35	–0.25	25	0	0.00	0.00
10–5	G	0.85	0.95	–0.35	–0.25	25/50	15/21	0.00	0.00
10–6	G	0.857	0.867	–0.270	–0.260	50	0	0.00	0.00
10–7	G	0.857	0.867	–0.270	–0.260	100	40	0.00	0.00
10–8	G	0.915	0.940	–0.315	–0.305	100	40	0.00	0.00
10–9	G	0.935	0.945	–0.305	–0.295	100	40	0.00	0.00
10–10	G	0.925	0.935	–0.295	–0.285	100	40	0.00	0.00
10–11	G	0.857	0.867	–0.270	–0.260	100	40	0.00	0.00
10–12	G	0.900	0.92	–0.255	–0.275	150	60	0.00	0.00
10–13	G	1.044	1.172	–0.299 2	–0.211 6	60	30	0.00	0.00
10–14	G	1.044	1.172	–0.299 2	–0.211 6	60	30	0.00	0.00
10–15	G	1.044	1.172	–0.299 2	–0.211 6	60	30	0.00	0.00
10–16	G	0.75	0.74	0.108	0.115 5	120	99	0.00	0.00
10–19	G	0.745 05	0.745 54	0.112 91	0.113 24	400	100	0.00	0.00
10–20	G	0.745 34	0.745 90	0.112 95	0.113 05	400	140	0.00	0.00
10–21	G	0.015 36	0.015 40	1.020 72	1.020 75	300	60	0.00	0.00
12.1–1	F	1.80	3.00	0.00	1.50	50	100	–	–
12.4–2	F	1.80	3.00	0.00	1.50	50	100	–	–

Table 13–1 Data for selected pictures.

13.2   Figure Index

13.3 Program Index

We list here both programs and program fragments (without distinguishing them). Each represents the algorithmic heart of the solution to some problem. By embedding these procedures in a surrounding program you obtain a runnable Pascal program. It is left to you as an exercise to declare the requisite global variables, to change the initialisation procedure appropriately, and to fit together the necessary fragments (see hints in Chapters 11 and 12). The heading 'Comments' states which problem the procedures form a solution to.

Program	Page	Comments
2.1–1	23	Measles numeric
2.1.1–1	29	Measles graphical
2.1.2–1	35	Graphical iteration
2.2–1	38	Feigenbaum iteration
2.2–2	42	Print-out of k in a running program
3.1–1	58	Verhulst attractor
3.2–1	63	Hénon attractor
3.3–1	66	Lorenz attractor
5.1–1	93	Assignment for iteration equation
5.1–2	94	Belongs to z_C
5.1–3	94	Mapping
5.1–4	95	JuliaNewtonComputeAndTest
5.1–5	96	StartVariableInitialisation
5.1–6	96	Compute
5.1–7	97	Test
5.1–8	97	Distinguish (does the point belong to the boundary?)
5.1–9	98	Distinguish (does the point belong to the basin?)
5.1–10	100	Distinguish (iteration number MOD 3 etc.)
5.2–1	109	Formulas in Pascal
5.2–2	113	JuliaNewtonComputeAndTest
5.2–3	122	Backwards iteration
5.2–4	123	Roots
6.1–1	135	MandelbrotComputeAndTest
6.3–1	159	Equality test for real numbers
6.3–2	159	Equality test for complex numbers
6.3–3	160	Mapping
6.3–4	164	Working part of Mapping
6.3–5	164	Drawing commands, real part
6.3–6	164	Drawing commands, imaginary part

Program	Page	Comments
12.4–2	354	Lightspeed Pascal Reference Program for Macintosh
12.5–1	361	ST Pascal Plus Reference Program for Atari
12.6–1	366	UCSD Pascal Reference Program for Apple II
12.6–2	367	Include–File of useful subroutines
12.6–3	371	TMLPascal Reference Program for Apple IIGS

13.4 Bibliography

H. Abelson and A. A. diSessa (1982). *Turtle Geometry*, MIT Press.

M. Abramowitz and J. A. Stegun (1968). *Handbook of Mathematical Functions*, Dover, New York.

Apple Computer Inc. (1980). *Apple Pascal Language Reference Manual* and *Apple Pascal System Operating Manual*.

K.-H. Becker and G. Lamprecht (1984). *PASCAL - Einführung in die Programmiersprache*, Vieweg.

R. Breuer (1985). Das Chaos, *GEO* 7 July 1985.

K. Clausberg (1986). Feigenbaum und Mandelbrot, Neue symmetrien zwischen Kunst und Naturwissenschaften, *Kunstform International* 85 (September/October 1986).

P. Collet, J.-P. Eckmann, and O. E. Lanford (1980). Universal properties of maps on an interval, *Communications in Mathematical Physics* 76 (1980) 211.

J. P. Crutchfield, J. D. Farmer, N. H. Packard, and R. S. Shaw, *et al.* (1986). Chaos, *Scientific American* (December 1986) 38-49.

J. H. Curry, L. Garnett, and D. Sullivan (1983). On the iteration of a rational function: computer experiments with Newton's method, *Commun. Math. Phys.* 91 (1983) 267-77.

U. Deker and H. Thomas (1983). Unberechenbares Spiel der Natur - Die Chaos-Theorie, *Bild der Wissenschaft* 1 (1983).

A.K.Dewdney (1986a). Computer recreations, *Scientific American* 255 (September 1986) 14-23.

A.K.Dewdney (1986b). Computer recreations, *Scientific American* 255 (December 1986) 14-18.

Werner Durandi (1987). Schnelle Apfelmännchen, *c't* 3 (1987).

Steve Estvanik (1985). From fractals to Graftals, *Computer Language*, March 1985, 45-8.

G. Y .Gardner (1985). Visual simulation of clouds, *Computer Graphics* 19 No.3.

Heimsoeth Software (1985). *Turbo Pascal 3.0 Manual*.

D. R. Hofstadter (1981). Metamagical Themas, *Scientific American* (November 1981) 16-29.

D. A. Huffmann [1952]. A method for the construction of minimum-redundancy codes, *Proc. Inst. Radio Engrs.* 40 (1952) 1698.

Gordon Hughes [1986]. Hénon mapping with Pascal, *Byte*, December 1986, 161.

S. Lovejoy and B. B. Mandelbrot (1985). Fractal properties of rain, and a fractal model, SIGGRAPH 85.

B. B. Mandelbrot (1977). *Fractals: Form, chance, and dimension*, Freeman, San Francisco.

P. Mann (1987). Datenkompression mit dem Huffmann Algorithmus, *User Magazin* (Newsletter of the Apple User Group, Europe) 22-3.

R. M. May (1976). Simple mathematical models with very complicated dynamics, *Nature* **261** (June 10 1976).

D. R. Morse, J. H. Lawton, M. M. Dodson, and M. H. Williamson (1985). Fractal dimension of vegetation and the distribution of arthropod body lengths, *Nature* **314**, 731-733.

M. M. Newman and R. F. Sproull (1979). *Principles of Interactive Computer Graphics*, McGraw-Hill.

H.-O. Peitgen and P. Richter (1985). Die undendliche Reise, *GEO* **7** (July 1985).

H.-O. Peitgen and P. Richter (1986). *The Beauty of Fractals*, Springer, Berlin, Heidelberg, New York.

H.-O. Peitgen and D. Saupe (1985). Fractal images - from mathematical experiments to fantastic shapes: dragonflies, scorpions, and imaginary planets, SIGGRAPH 85.

Research Group in Complex Dynamics, Bremen (1984a). *Harmonie in Chaos und Cosmos*, Exhibition catalogue, Sparkasse Bremen, 16.1.84-3.2.84.

Research Group in Complex Dynamics, Bremen (1984b). *Morphologie Komplexer Grenzen*, Exhibition catalogue, Max Planck Institute for Biophysical Chemistry 27.5.84-9.6.84; Exhibition catalogue, Sparkasse Bonn 19.6.84-10.7.84.

Research Group in Complex Dynamics, Bremen (1985). *Schönheit im Chaos; Computer Graphics face complex dynamics*. Exhibition catalogue, Goethe Institute, Munich.

W. Rose (1985). Die Entdeckung des Chaos, *Die Zeit* **3** (11 January 1985).

A. M. Saperstein (1984). Chaos - a model for the outbreak of war, *Nature* (24 May 1984).

SIGGRAPH (1985) Special interest group on computer graphics. Course notes: *Fractals, Basic Concepts, Computation and Rendering*.

Alvy Ray Smith (1984). Plants, fractals, and formal languages, *Computer Graphics* **18** No. 3 (July 1984).

Frank Streichert (1987). Informationverschwendung - Nein danke: Datenkompression durch Huffmann-Kodierung, *c't* **1** (January 1987).

E. Teiwes (1985). *Programmentwicklung in UCSD Pascal*, Vogel, Würzburg.

R. Walgate (1985). Je kleiner das Wesen, desto grösser die Welt, *Die Zeit* **20** (10 May 1985).

N.Wirth (1983). *Algorithmen and Datenstrukturen*, Teubner, Stuttgart.

13.5   Acknowledgements

We are grateful to all chaos researchers who have helped us with information, discussions, and criticisms:

Dipl.-Phys. Wolfram Böck, Bremen
Dr Axel Brunngraber, Hanover
Prof. Dr. Helmut Emde, TH Darmstadt
Dr Hartmut Jürgens, Bremen
Dipl.-Inform. Roland Meier, Bremen

Index